FRONTIERS OF SCIENCE

WEATHER AND CLIMATE

FRONTIERS OF SCIENCE

WEATHER
AND CLIMATE

Notable Research and Discoveries

KYLE KIRKLAND, PH.D.

Facts On File
An imprint of Infobase Publishing

WEATHER AND CLIMATE: Notable Research and Discoveries

Facts On File, Inc.
An imprint of Infobase Publishing
132 West 31st Street
New York NY 10001

Library of Congress Cataloging-in-Publication Data

Kirkland, Kyle.
  Weather and climate : notable research and discoveries / Kyle Kirkland.
      p. cm. — (Frontiers of science)
  Includes bibliographical references and index.
  ISBN 978-0-8160-7446-4
  1. Climatology. 2. Weather. 3. Discoveries in science. I. Title.
  QC861.3.K57 2010
  551.5—dc22   2009028193

Facts On File books are available at special discounts when purchased in bulk quantities for businesses, associations, institutions, or sales promotions. Please call our Special Sales Department in New York at (212) 967-8800 or (800) 322-8755.

You can find Facts On File on the World Wide Web at http://www.factsonfile.com

Text design by Kerry Casey
Illustrations by Richard Garratt and Infobase
Photo research by Tobi Zausner, Ph.D.
Composition by Mary Susan Ryan-Flynn
Cover printed by Bang Printing, Inc., Brainerd, Minn.
Book printed and bound by Bang Printing, Inc., Brainerd, Minn.
Date printed: June 2010
Printed in the United States of America

10 9 8 7 6 5 4 3 2 1

CONTENTS

PREFACE

Discovering what lies behind a hill or beyond a neighborhood can be as simple as taking a short walk. But curiosity and the urge to make new discoveries usually require people to undertake journeys much more adventuresome than a short walk, and scientists often study realms far removed from everyday observation—sometimes even beyond the present means of travel or vision. Polish astronomer Nicolaus Copernicus's (1473–1543) heliocentric (Sun-centered) model of the solar system, published in 1543, ushered in the modern age of astronomy more than 400 years before the first rocket escaped Earth's gravity. Scientists today probe the tiny domain of atoms, pilot submersibles into marine trenches far beneath the waves, and analyze processes occurring deep within stars.

Many of the newest areas of scientific research involve objects or places that are not easily accessible, if at all. These objects may be trillions of miles away, such as the newly discovered planetary systems, or they may be as close as inside a person's head; the brain, a delicate organ encased and protected by the skull, has frustrated many of the best efforts of biologists until recently. The subject of interest may not be at a vast distance or concealed by a protective covering, but instead it may be removed in terms of time. For example, people need to learn about the evolution of Earth's weather and climate in order to understand the changes taking place today, yet no one can revisit the past.

Frontiers of Science is an eight-volume set that explores topics at the forefront of research in the following sciences:

- biological sciences
- chemistry

- computer science
- Earth science
- marine science
- physics
- space and astronomy
- weather and climate

The set focuses on the methods and imagination of people who are pushing the boundaries of science by investigating subjects that are not readily observable or are otherwise cloaked in mystery. Each volume includes six topics, one per chapter, and each chapter has the same format and structure. The chapter provides a chronology of the topic and establishes its scientific and social relevance, discusses the critical questions and the research techniques designed to answer these questions, describes what scientists have learned and may learn in the future, highlights the technological applications of this knowledge, and makes recommendations for further reading. The topics cover a broad spectrum of the science, from issues that are making headlines to ones that are not as yet well known. Each chapter can be read independently; some overlap among chapters of the same volume is unavoidable, so a small amount of repetition is necessary for each chapter to stand alone. But the repetition is minimal, and cross-references are used as appropriate.

Scientific inquiry demands a number of skills. The National Committee on Science Education Standards and Assessment and the National Research Council, in addition to other organizations such as the National Science Teachers Association, have stressed the training and development of these skills. Science students must learn how to raise important questions, design the tools or experiments necessary to answer these questions, apply models in explaining the results and revise the model as needed, be alert to alternative explanations, and construct and analyze arguments for and against competing models.

Progress in science often involves deciding which competing theory, model, or viewpoint provides the best explanation. For example, a major issue in biology for many decades was determining if the brain functions as a whole (the holistic model) or if parts of the brain carry out specialized functions (functional localization). Recent developments in brain imaging resolved part of this issue in favor of functional localization by showing that specific regions of the brain are more active during certain tasks. At the same time, however, these experiments have raised other questions that future research must answer.

The logic and precision of science are elegant, but applying scientific skills can be daunting at first. The goals of the Frontiers of Science set are to explain how scientists tackle difficult research issues and to describe recent advances made in these fields. Understanding the science behind the advances is critical because sometimes new knowledge and theories seem unbelievable until the underlying methods become clear. Consider the following examples. Some scientists have claimed that the last few years are the warmest in the past 500 or even 1,000 years, but reliable temperature records date only from about 1850. Geologists talk of volcano hot spots and plumes of abnormally hot rock rising through deep channels, although no one has drilled more than a few miles below the surface. Teams of neuroscientists—scientists who study the brain—display images of the activity of the brain as a person dreams, yet the subject's skull has not been breached. Scientists often debate the validity of new experiments and theories, and a proper evaluation requires an understanding of the reasoning and technology that support or refute the arguments.

Curiosity about how scientists came to know what they do—and why they are convinced that their beliefs are true—has always motivated me to study not just the facts and theories but also the reasons why these are true (or at least believed). I could never accept unsupported statements or confine my attention to one scientific discipline. When I was young, I learned many things from my father, a physicist who specialized in engineering mechanics, and my mother, a mathematician and computer systems analyst. And from an archaeologist who lived down the street, I learned one of the reasons why people believe Earth has evolved and changed—he took me to a field where we found marine fossils such as shark's teeth, which backed his claim that this area had once been under water! After studying electronics while I was in the air force, I attended college, switching my major a number of times until becoming captivated with a subject that was itself a melding of two disciplines—biological psychology. I went on to earn a doctorate in neuroscience, studying under physicists, computer scientists, chemists, anatomists, geneticists, physiologists, and mathematicians. My broad interests and background have served me well as a science writer, giving me the confidence, or perhaps I should say chutzpah, to write a set of books on such a vast array of topics.

Seekers of knowledge satisfy their curiosity about how the world and its organisms work, but the applications of science are not limited to intellectual achievement. The topics in Frontiers of Science affect society on a multitude of levels. Civilization has always faced an uphill battle to procure scarce resources, solve technical problems, and maintain order. In modern times, one of the most important resources is energy, and the physics of fusion potentially offers a nearly boundless supply. Technology makes life easier and solves many of today's problems, and nanotechnology may extend the range of devices into extremely small sizes. Protecting one's personal information in transactions conducted via the Internet is a crucial application of computer science.

But the scope of science today is so vast that no set of eight volumes can hope to cover all of the frontiers. The chapters in Frontiers of Science span a broad range of each science but could not possibly be exhaustive. Selectivity was painful (and editorially enforced) but necessary, and in my opinion, the choices are diverse and reflect current trends. The same is true for the subjects within each chapter—a lot of fascinating research did not get mentioned, not because it is unimportant, but because there was no room to do it justice.

Extending the limits of knowledge relies on basic science skills as well as ingenuity in asking and answering the right questions. The 48 topics discussed in these books are not straightforward laboratory exercises but complex, gritty research problems at the frontiers of science. Exploring uncharted territory presents exceptional challenges but also offers equally impressive rewards, whether the motivation is to solve a practical problem or to gain a better understanding of human nature. If this set encourages some of its readers to plunge into a scientific frontier and conquer a few of its unknowns, the books will be worth all the effort required to produce them.

ACKNOWLEDGMENTS

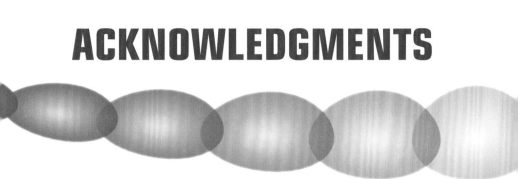

Thanks go to Frank K. Darmstadt, executive editor at Facts On File, and the rest of the Facts On File staff for all their hard work, which I admit I sometimes made a little bit harder. Thanks also to Tobi Zausner for researching and locating so many great photographs. I also appreciate the time and effort of a large number of researchers who were kind enough to pass along a research paper or help me track down some information.

INTRODUCTION

One of the best "tests" of knowledge is the ability to make accurate predictions with it. Tests constitute the framework of the scientific method—ideas or prior research generate hypotheses, which make predictions that can be tested and supported (or rejected) by experiment. This method is vital for weather and climate science, and it also plays a critical role in the most widespread application of this science—forecasting. Forecasting the path of a hurricane and Earth's future climate involve a large number of tests as forecasters study patterns, analyze data, and make predictions to be compared with actual results. Any failures mean that the forecasting techniques need refining.

Variability in weather and climate makes the study of these phenomena interesting and is also the main reason why forecasting is important—changes that come tomorrow, next year, or 50 years from now can have tremendous impacts. Weather and climate are related topics but taken individually are not quite the same thing. The main difference involves time—weather refers to the present state or to conditions that may evolve in the coming days or weeks, while climate refers to long-term behavior. Both weather and climate are functions of the atmosphere—the air surrounding the planet—and its interactions with land and water surfaces. *Meteorology* is the study of atmospheric phenomena and processes such as weather patterns. The term *meteorology* derives from Greek words *meteoron,* meaning high in the air, and *logos,* meaning knowledge or word. (Meteorology should not be confused with meteoritics, the study of meteors, which name also derives from the Greek *meteoron.*)

Weather and Climate, one volume in the Frontiers of Science set, is about explorers and scientists who delve into frontiers of weather and

This image, taken by the *GOES-8* weather satellite, shows the clouds over North and South America. *(NASA/Goddard Space Flight Center)*

climate science—and quite often run into variable and puzzling phenomena. Earth's atmosphere and its interactions form extensive, complicated systems, and Earth is home to climates as different as those of rain forests and deserts. Predicting the behavior of these systems is exceptionally difficult because the systems are highly sensitive to outside conditions—a slight change in temperature or pressure, for example, may result in a huge change in behavior. As a result, weather forecasts that are issued more than a few days in advance are often wrong.

Atmospheric phenomena affect everyone—severe storms pop up everywhere from time to time, and the interactions governing the climate form a network with worldwide influence. These topics are complicated and merit a great deal of research, and each chapter of this book explores one of these frontiers. Reports published in journals, presented at conferences, and described in press releases illustrate the kind of research problems of interest in weather and climate science and how scientists attempt to solve them. *Weather and Climate* summarizes a selection of these reports—unfortunately there is room for only a fraction of them—that offers students and other readers insights into the methods and applications of meteorology and climatology.

Students need to keep up with the latest developments in these quickly moving fields of research, but they have difficulty finding a source that explains the basic concepts while discussing the background and context essential for the big picture. This book describes the evolution of each of the six main topics it covers and explains the problems that researchers are currently investigating as well as the methods they are developing to solve them.

Three of the chapters focus on climate and are specifically devoted to change and variation. Average surface temperature is increasing on a global scale, and scientists are rapidly gaining knowledge of *climate forcings*—mechanisms that cause or force the climate to change. An understanding of forcings and the reasons for the current changes will help scientists predict the future course of climate change, which is essential in order to take appropriate action to eliminate or reduce contributing factors. But to understand the evolution of Earth's climate, researchers must know something about the past. Paleoclimatology, the subject of chapter 1, is the study of the past climate. Knowledge of Earth's past climate is a great help in determining what may happen in the future.

Chapter 2 focuses on climate change at the poles. The climate around the North and South Poles is harsh and frigid, but these environments form an important component of the world's weather and climate. Global warming has generally resulted in rapid losses of ice and snow in these areas, the consequences of which are the focus of much research. Polar climates are highly susceptible to change, especially the Arctic, and the findings of research conducted in this region could be a harbinger of things to come elsewhere on Earth.

Much of the research on global climate change points to the increase in particular emissions called *greenhouse gases,* which are generated and released in a number of industrial and technological processes. Other factors may also be contributing. Although the role of these factors seems far too small to account for most of the recent warming trend, their effects over the long term are less certain. Chapter 3 describes research on variations in the Sun and the radiation that bathes Earth. Most of the energy that drives the planet's weather and climate comes from solar radiation, which makes variability in the Sun's output an important subject—and potentially a factor in Earth's climate.

The final three chapters focus on violent weather systems or ways to modify and control various aspects of the weather. Chapter 4 looks at tornadoes—these violent, short-lived phenomena can be found everywhere, but many of the storms, and almost all of the most powerful ones, occur in the United States, causing dozens of fatalities and millions of dollars worth of damage each year. Meteorologists are making progress in spotting these storms in their earliest stages and issuing accurate warnings, but the formation and development of tornadoes remain mysterious. Storm chasers—researchers who drive heavily instrumented vehicles and collect data in the field—as well as improving radar instruments are spearheading the effort to learn more.

Chapter 5 examines hurricane forecasting. Before the era of satellites, which began in the latter half of the 20th century, scientists who tracked and studied these marine storms had to rely on spotty observations by ships at sea and from islands in the path of the storm. Today, researchers have many tools to use, and intrepid aviators fly into the storms all the way to the eye—the calm center of the hurricane—to study how hurricanes form, evolve, and move. Forecasting accuracy has improved recently, but there is much more research left to do concerning the path that hurricanes take and what fuels their intensity.

Stormfury, an ambitious project that was active in the 1960s and early 1970s, studied the possibility of developing techniques to weaken hurricanes before they came ashore. Although the project's results were inconclusive, other efforts to modify the weather have since become popular. Several of the drier states in the western United States engage in rainmaking operations and have reported some success. But these reports are controversial because of the difficulty in determining the effects of weather modification—no one knows what would have happened in the absence of these operations, which means gauging the success or failure of the operations is complicated. Chapter 6 discusses research that may resolve the issue once and for all.

Meteorology and climatology deal with large and complex subjects, and research in these branches of science is sometimes contentious. The importance of these subjects to everyone on the planet means that billions of people have a direct stake in the findings—and an opinion on what should be done and how the results should be interpreted. These opinions often vary, and this book attempts to cover all sides of the issue, though the consensus, if there is one, is highlighted. But opinions will likely shift and change almost as much as the weather does, because investigators at the frontiers of science will continue to make new discoveries in these fascinating and crucial fields of research.

Paleoclimatology— Evolution of Earth's Climate

The Greek philosopher Heraclitus (ca. 535–475 B.C.E.) highlighted the importance of change in his views of the world. Consider a river—although it may seem to be unchanging, it is constantly flowing, discharging its contents into the ocean and renewing itself with springs, rain runoff, and snowmelt. Earth's climate can also appear static, fluctuating with the seasons but remaining consistent and unchanging in its patterns. But paleoclimatology—the study of past climate—shows that the planet's climate has not stayed the same throughout its history. Earth is dynamic, not static.

Climate change has made the headlines recently. According to the Intergovernmental Panel on Climate Change (IPCC), an international body of scientists and government officials established by the United Nations Environment Programme (UNEP) and the World Meteorological Organization (WMO), the average surface temperature on Earth has risen 1.33°F (0.74°C) in the last 100 years. Associated with rising temperatures are hazards such as melting *glaciers,* which could elevate sea levels and inundate low-lying coastlines and islands, and potential disruptions in weather patterns and rainfall. The National Research Council, one of the National Academies that address scientific issues and provide advice to the U.S. government, announced in 2006 that they had "a high level of

Yahtse Glacier at Icy Bay, Alaska *(Captain Budd Christman, NOAA Corps)*

confidence that the last few decades of the 20th century were warmer than any comparable period in the last 400 years," according to a National Academies news release dated June 22, 2006.

Such statements are startling, yet they also invite skepticism. Reliable weather records only go back to the middle of the 19th century, so scientists must use other sources to extend their historical knowledge. These sources are the tools and techniques of paleoclimatology. Researchers gather *proxy climate data,* written in coral reefs, tree growth, glaciers, ice, and sediment deposits, to piece together the history of Earth's climate.

Climate change happens, and no one is sure exactly why these changes occur or what the future might bring. Because weather and climate affect economies, crop yields, and so many other aspects of society, these issues must be addressed. Researchers are also studying the alteration of the planet and its atmosphere due to increasing pollution and other emissions. To get an idea of what is coming, scientists rely on the study of the past, which enables the development of climate models capable of making accurate predictions. This chapter describes the methods and concepts of paleoclimatology and discusses how this frontier of science is helping people understand—and potentially preserve—the global climate.

INTRODUCTION

Earth in its earliest stages looked much different than it does today. Scientists believe that the Sun and planets formed about 4.5 billion years ago from a swirling cloud of gas and dust. Radioactive *isotopes* bear witness to Earth's great age; the nuclei of these atoms decay at a constant rate, and researchers use the population ratio of certain nuclei as a clock to measure the lapse of time. (Carbon isotopes are useful in dating recent events, as described in the sidebar on page 115.) Astronomers have observed distant star systems forming out of clouds of gas and dust, similar to the process that took place billions of years ago for the Sun and its retinue of planets.

Planets gradually form as the force of gravitation brings clumps of matter together. After its birth, Earth consisted of molten rock, heated with the energy of the Sun and the violent impacts of stray debris. Much of the gas on the planet would have been hydrogen, which is the lightest and most common element in the universe. But the high temperature of the early planet would have driven away most of this element. This is because high temperatures correspond with greater atomic motion, and the heated, lightweight hydrogen gas would have had enough energy to escape Earth's gravity. Earth gradually cooled as it radiated some of its heat energy into space.

Two different sources may have contributed to Earth's early atmosphere—one from the planet's interior and one from space. Much of the planet's early atmosphere probably came from volcanic activity. Volcanoes must have been common in this hot, early period, and, in addition to lava, their emissions would have included gases that had been trapped in the interior as the planet formed. The following figure illustrates the release of prominent gases into the atmosphere. Hydrogen, nitrogen, carbon dioxide (CO_2), and *water vapor* poured onto the surface.

Bombardment from space may have also contributed to the early atmosphere. Rocks and bits of matter left over from the solar system's

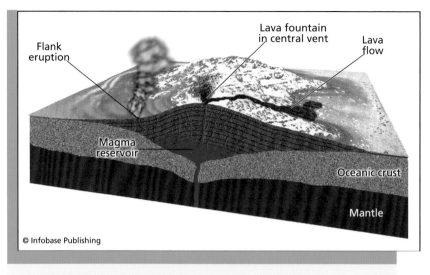

© Infobase Publishing

Volcanoes emit gases as well as ash into the atmosphere.

creation would have bombarded the young planet—the scars of these impacts can been seen on the Moon and other airless worlds that have no weather to erase craters. Some of these impacts could have been comets, bringing frozen water, CO_2, and other compounds that transitioned to the gaseous phase in the warm environment.

Oxygen was a latecomer to Earth's atmosphere. Fossil evidence indicates that life arose quickly, and the metabolic by-products of these organisms included oxygen, as is true for modern plants. The composition of Earth's atmosphere today by volume is 78 percent nitrogen, 21 percent oxygen, and a number of other gases, including a small amount of CO_2 and water vapor.

The atmosphere is critically important in Earth's climate. Earth receives most of its energy input in the form of sunlight, which consists of *electromagnetic radiation* of various frequencies. The planet absorbs sunlight, raising its temperature. Winds and ocean currents distribute this heat so that there is not a huge difference between temperatures during the day and night. The temperature drops at night but not much. In Atlanta, Georgia, for example, the high temperature of the day averages about 20–25°F (11.1–13.8°C) above the low temperature. Compare that change to Mars, which has a thin atmosphere having less than 1 percent of the pressure of Earth's atmosphere. Probes of the *National Aeronautics and Space Administration (NASA)* have landed on Mars and reported average surface temperatures reach -20°F (-29°C) during the day and plunge to -120°F (-84°C) at night—a much greater range than on Earth, even though the Martian day is only slightly longer.

Composition of the atmosphere is another crucial factor. Certain atmospheric gases have a greater or lesser ability to absorb specific portions of the electromagnetic spectrum, which has a strong effect on Earth's temperature. Gases such as oxygen (most of which exists in the atmosphere as a molecule, O_2) and nitrogen (most of which is N_2) do not absorb visible light, so Earth's atmosphere is highly transparent to this portion of the spectrum. Some of this light reflects from Earth's surface, but part of it gets absorbed, raising the temperature of the ground, ocean, or object that absorbed it. These warmed objects reemit the radiation, which is one of the heat transfer mechanisms by which they cool off. But they do not have enough energy—or in other words, their temperature is not high enough—to release much of it in the form of visible light. Instead, warm objects emit a lot of infrared radiation, which has a lower frequency—and therefore lower energy—than visible light.

Gases such as CO_2, water vapor, methane (CH_4), and others absorb infrared radiation. They tend to block the emissions from warm objects, trapping the heat. The effect is similar to a greenhouse, in which clear panes of glass allow light to enter but block most of the infrared radiation. Gases that absorb a lot of infrared radiation are known as greenhouse gases.

Greenhouse gases in the atmosphere are important in keeping the Earth warm and comfortable, but a change in the amount of these gases may lead to serious consequences. Emissions related to modern technology and industry often include a significant percentage of greenhouse gases. People worry that drastic climate changes will occur if the amount of these gases continues to rise. But it is extremely difficult to understand a system as complex as Earth's climate well enough to make predictions about its future behavior. One of the most important motivations for the study of past climate is the insight it gives into what the climate might become in the future.

RECONSTRUCTING ANCIENT CLIMATES

Although meteorologists have only been documenting the weather for about a century and a half, the effects of past events are written in the Earth and its features. Geology is the study of Earth and its history, and geologists do much of their work by collecting and analyzing rocks. Different kinds of rocks have difference sources—igneous rocks form when magma (molten rock) cools, and sedimentary rocks are created when *sediments* such as mud or the calcium carbonate shells of certain marine organisms get buried and compacted. These rocks tell a story that geologists read by studying their location, composition, and age.

Sediments and sedimentary rocks are particularly important in paleoclimatology because in many cases they form at the bottom of seas. Water is a critical agent in climate; oceans hold and circulate a great deal of heat, and *precipitation* is a major part of weather. The three phases of water—liquid, gas (water vapor), and solid (ice)—participate in the *water cycle,* as water evaporates from the ocean and falls as rain or snow, most of which eventually returns to the ocean by draining into rivers and falling into the sea, to start the cycle again. Sediments in water tend to drop to the bottom and collect over time. The heavy weight of the water above can compress the sediments into rocks, given sufficient time.

Ocean sediments yield much information about the state of the ocean at the time the sediment was deposited. For example, sediment composition can tell scientists about how the level and amount of water in the seas have risen and fallen. The total amount of water on Earth tends to remain constant, but the proportion of each phase can vary—when one phase increases, the others must decrease. In cold spells, water tends to get trapped in ice, so there is less available for the seas. Warm weather melts the ice, and bodies of water swell.

The ratio of two specific oxygen isotopes is one of the most important techniques researchers use to study past sea levels. Isotopes generally have the same chemical properties, but their difference in mass—due to the varying number of *neutrons* in the nucleus—distinguishes them. Oxygen-16 is the most common oxygen isotope, containing eight *protons* and eight neutrons (for a total of 16 particles in the nucleus). This isotope is stable—it does not exhibit radioactive decay. Oxygen-18 is another stable isotope that has 10 neutrons (and eight protons, which is true for all oxygen atoms) and is therefore heavier than oxygen-16. Water consists of two atoms of hydrogen and one atom of oxygen (H_2O), and if the oxygen atom is oxygen-18, the water molecule weighs

National Oceanic and Atmospheric Administration (NOAA)

In 1970, the U.S. government grouped a collection of agencies under one roof, leading the National Oceanic and Atmospheric Administration (NOAA). Although formed in 1970, some of the agencies that are part of it are much older—for example, the U.S. Coast and Geodetic Survey, created in 1807, and the Weather Bureau, formed in 1870. Today, NOAA includes many organizations, such as the National Weather Service (NWS), the Office of Oceanic and Atmospheric Research (OAR), the National Marine Fisheries Service (NMFS), and others.

According to its Web site, NOAA's mission is as follows: "To understand and predict changes in Earth's environment and conserve and manage coastal and marine resources to meet our Nation's economic, social, and environmental needs." The Earth's atmosphere and oceans play critical roles in weather and climate, so NOAA is greatly involved in meteorology and climate studies, as evidenced by the inclusion of organizations such as the NWS, which issues forecasts and tracks storms, providing much-needed warnings when threatening systems such as hurricanes develop.

One of the most important NOAA components in terms of climatology is the National Climatic Data Center (NCDC), the largest weather and climate data archive in the world. Maintaining most of the records at an office in Asheville, North Carolina, the center acquires, processes, summarizes, maintains, and publishes millions of paper and digital data records. The data include weather charts, paleoclimatology research findings, satellite observations, and much else.

more than if it is oxygen-16. "Heavier" water tends to sink and has less of chance of evaporating. Because oxygen-16 is lighter, it is the quickest to leave, and during cold periods, when a lot of water gets trapped in ice, the proportion of oxygen-16 drops in the ocean. The ratio of these oxygen isotopes in ocean sediments gives researchers a clue about the sea level at the time of the deposit.

Other clues come from fossils. Some animals, such as the extinct woolly mammoth, thrived in cold environments. Studying the location and ages of the fossils of these creatures provides information on the state of the climate when the animal was alive.

This data must be collected and archived. One of the most important agencies responsible for maintaining these collections is the *National Oceanic and Atmospheric Administration (NOAA),* part of the U.S. Department of Commerce. The sidebar on page 7 provides more information on this large agency.

Techniques to gather and study proxy climate data will be discussed in more detail in the following sections. Researchers are continually refining these techniques to extract increasing amounts of valuable information. However, a broad sketch of Earth's past climate has emerged. As described in NOAA's "Introduction to Paleoclimatology," posted at their Web site, "If there is one thing that the paleoclimatic record shows, it is that the Earth's climate is always changing."

One of the most prominent changes is warming or cooling trends. Many periods throughout Earth's history have witnessed extremes in warm or cold weather, with the cold spells called ice ages. An ice age occurs when a great deal of water ends up in the solid state, accumulating in large masses of ice called glaciers that cover much of the land surface. Research on oxygen isotope ratios and other studies indicate a number of major and minor ice ages, the last of which began about 100,000 years ago and reached its coolest temperature about 20,000 years ago, finally ending about 12,000 years ago.

Pinpointing the causes and factors of these trends is crucial for understanding and predicting climate. The Serbian scientist Milutin Milanković (also known as Milankovitch) identified one of the factors involved in a 100,000-year cycle. In the 1920s, Milanković proposed that cyclical variations in Earth's orbit about the Sun reduce the amount of radiation the planet receives during certain periods, lowering global temperatures. Earth's past climate indicates that this cycle is operating on 100,000-year periods—the last few ice ages have been spaced about 100,000 years apart—but the variations appear too weak to explain all of the climate changes. (See chapter 3 for more information on the effects of solar radiation on Earth.)

Other fluctuations covering longer or shorter periods are also not yet fully understood. But researchers are beginning to get some ideas. In "The Big Chill," an article posted at the Public Broadcasting Service's (PBS) NOVA site online, the University of Maine researcher Kirk A. Maasch wrote, "Although the exact causes for ice ages, and the glacial cycles within them, have not been proven, they are most likely the result of a complicated dynamic interaction between such things as solar output, distance of the Earth from the Sun, position and height of the continents, ocean circulation, and the composition of the atmosphere."

WRITTEN IN THE ICE

To read the past climate's record that has been written in the Earth's features, climatologists travel to desolate, undisturbed places. Some of the most important paleoclimatology data comes from the vast ice sheets of Greenland and Antarctica.

Greenland is the world's largest island, about three times as large as the state of Texas, and is situated between the Arctic Ocean and the North Atlantic Ocean. Most of Greenland is covered in ice, which means that Greenland is somewhat misnamed, though the southern coasts, where early Scandinavian colonists probably settled, are green with vegetation in the summer. Antarctica is a continent about one and a half times the size of the United States and is located at the southern end of the world, at and around the South Pole. Its name derives from a Greek word *antarktikos,* meaning the opposite of north. Ice blankets almost the entire cold and forbidding continent.

Snow falls in many parts of Greenland and Antarctica throughout the year. Because of the continually frigid conditions, most of this snow does not melt but instead accumulates in layers, with the lower layers being older. The pressure of the weight from new layers compresses the lower, older layers into dense ice sheets. In many cases the ice forms a glacier—a large mass of ice that gravity slowly pulls down mountain valleys or across plains that slope toward the ocean. Some glaciers are only the size of a football field, but others can be many times as large.

However, some of the ice is stable and stays in one place. In these locations, the ice accumulates over a period of thousands of years. In parts of Antarctica, the ice is two miles (3.2 km) thick or more. Buried within these ice sheets are hints of what the climate was like at the time

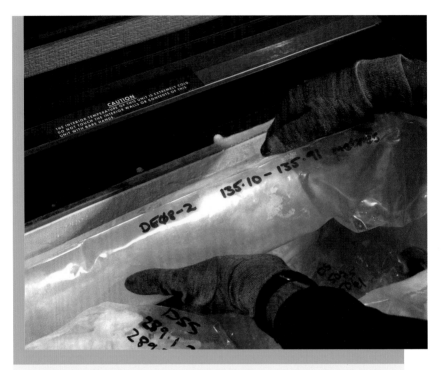

A researcher studies an ice core at the Ice Core Extraction Laboratory in Melbourne, Australia. *(Simon Fraser/CSIRO/Photo Researchers, Inc.)*

the snow fell. The ice contains thousands of years of history. To reach the data, scientists cut into the ice with a hollow drill and extract a long, cylindrical core.

As the writer Holli Riebeek discussed in "Paleoclimatology: The Ice Core Record," the properties of ice cores are proxies—representatives—of the climate conditions at the time they were laid down. "The ice cores can provide an annual record of temperature, precipitation, atmospheric composition, volcanic activity, and wind patterns. In a general sense, the thickness of each annual layer tells how much snow accumulated at that location during the year. Differences in cores taken from the same area can reveal local wind patterns by showing where the snow drifted. More importantly, the make-up of the snow itself can tell scientists about past temperatures."

Measurements of oxygen isotope ratios help researchers to reconstruct sea levels and temperatures of the past. The temperature of the ice core itself is also informative. Although the Sun and warm air heats the surface of the ice sheet, the buried material retains some of its original characteristics. "When scientists lower an ultra-precise thermometer into a hole in the ice," Riebeek wrote, "they can detect the temperature variations that have occurred since the Ice Age. The near-surface ice temperature, like the atmosphere today, is warm, and then the temperature drops in the layers formed roughly between A.D. 1450 and 1850, a period known as the Little Ice Age, one of several cold snaps that briefly interrupted the overall warming trend ongoing since the end of the [last] Ice Age."

An international team of scientists recently obtained one of the longest continuous ice cores from a plateau in the eastern part of Antarctica. The 1.8-mile (3-km) core contains 740,000 years of data. A press release posted on June 11, 2004, at Science*Daily* noted that the initial inspection agreed with previously obtained data. "The first results confirm that over the last 740,000 years the Earth experienced eight ice ages, when Earth's climate was much colder than today, and eight warmer periods (interglacials). In the last 400,000 years the warm periods have had a temperature similar to that of today. Before that time they were less warm, but lasted longer."

Ice cores are also valuable tools to study the relationship between climate conditions and the state of the atmosphere. This is because dust, pollen, and other particles, along with air, get trapped in the snow as it falls. The subsequent compression seals off little pockets containing important samples of the past atmosphere. A careful inspection of the contents of these tiny bubbles allows researchers to study the ancient atmosphere.

STUDYING GAS TRAPPED IN ICE

One recent study by the Oregon State University researchers Jinho Ahn and Edward J. Brook examined CO_2 variations in Antarctic ice cores. The samples in these cores ranged in age from 20,000 to 90,000 years. To get at the trapped air, the researchers crushed the ice and isolated the released gases. Delicate measurements indicated the level of this greenhouse gas in the samples. Ahn and Brook tested the levels at 1,000-year intervals. The National Science Foundation (NSF) funded this study, which was published in a 2008 issue of *Science*.

After measuring the amount of CO_2, the researchers compared the results at each time interval with the corresponding temperature, as obtained in proxy climate data from Greenland and Antarctica. Ahn and Brook also studied marine sediments in order to correlate the CO_2 data with oceanic conditions prevailing at the time.

An NSF press release issued on October 3, 2008, described the findings. "The researchers discovered that elevations in carbon dioxide levels were related to subsequent increases in the Earth's temperature as well as reduced circulation of ocean currents in the North Atlantic. The data also suggests that carbon dioxide levels increased along with the weakening of mixing of waters in the Southern Ocean. This, the researchers say, may point to a potential future scenario where global warming causes changes in ocean currents which in turn cause more carbon dioxide to enter the atmosphere, adding more greenhouse gas to an already warming climate."

This scenario is particularly dangerous because it is a positive feedback loop. Feedback occurs when the output of a process or system affects its function. Negative feedback arises when the output tends to turn down or diminish the system's activity; for example, engineers often use negative feedback loops in electrical circuits to keep the current from rising too much—an increase in current will decrease the system's activity, causing the current to fall back to an appropriate level. In contrast, positive feedback loops are dangerous because they ramp up systems. Positive feedback can cause runaway excitation as an increase in the output leads to a further increase, and so on.

A future in which positive feedback increases an already perilous trend is alarming. But the researchers note that further studies are necessary before any firm conclusions can be reached. More samples, measured with instruments having greater precision, of longer time periods of past climate will be essential.

TREE RINGS AND POLLEN

Excellent standards of precision can be achieved in some of the more recent periods when researchers study biological proxies. Weather and climate strongly affect an organism's environment, resulting in variable conditions in which the organism must live. For example, sometimes the temperature will be optimal and sometimes too cold or too hot;

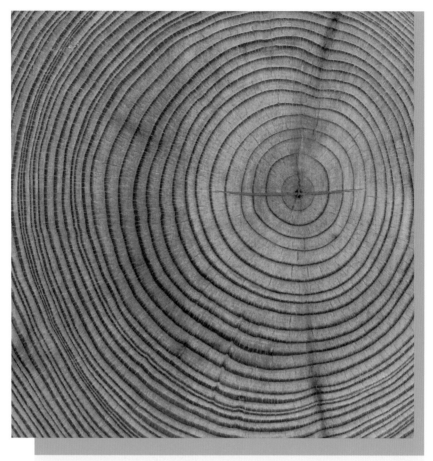

Tree rings *(Zastol'skiy Victor Leonidovich/Shutterstock)*

sometimes there will be enough rainfall and at other times too much or too little. The growth and population of certain organisms reflect the conditions they have experienced.

Trees are a great source of information on seasonal variation in rainfall. As trees age, their girth increases—the trunk becomes thicker. Each year adds another ring, as illustrated in the figure above. Rings are easily observed in the stump of a felled tree, and counting the number of rings in the stump reveals the age of the tree. Dendrochronology is the study of tree rings, deriving from the Greek words *dendros,* meaning tree, *chronos,* meaning time, and *logos,* word or study. The use of trees in the study of climate is called *dendroclimatology.*

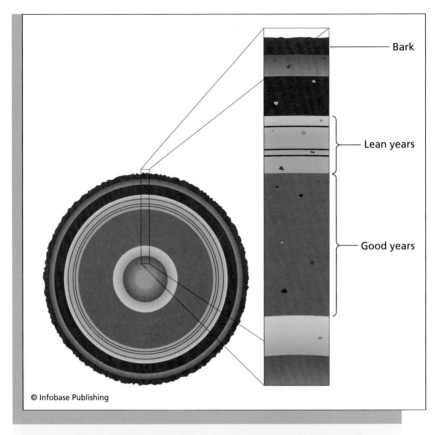

Bark

Lean years

Good years

© Infobase Publishing

Close inspection of this tree's rings shows some years with excellent growth—the wide rings—as well as the stunted growth of years in which the tree probably received little rain.

The width of tree rings indicates how much the tree grew during the year. Growth is strongly related to the amount of moisture the tree absorbed, which is an indication of rainfall. Some trees live thousands of years, providing dendroclimatologists with a valuable record of seasonal precipitation. A candidate for the world's oldest living tree is a 9,550-year-old spruce in the Dalarna province of Sweden. The trunk is not that old, but scientists dated root samples of the tree with radiocarbon dating. As described in the following sidebar, radiocarbon dating is an important technique to determine the age of organic material. The samples suggest that some portion of the tree's roots got their start 9,550 years ago, although most of the rest of the tree consists of recent additions or replacements for lost parts.

Radiocarbon Dating

Carbon is an essential element in all life on Earth. This element forms the backbone of long, complex molecules such as carbohydrates and many others that provide nutrients or perform metabolic or structural functions in cells. The most common carbon isotope is carbon-12, a stable isotope comprising about 99 percent of carbon atoms. Carbon-14 is a radioactive isotope with a *half-life* of 5,730 years—half of a sample of carbon-14 atoms decays every 5,730 years into nitrogen-14 when a proton in the nucleus changes into a neutron, emitting an *electron* and a neutral particle in the process. After 5,730 years, one-half of an original sample will have turned into nitrogen-14; after 11,460 (two half-lives), only one-quarter (half of a half) of the carbon-14 atoms will remain, and so on, until most of the carbon has decayed.

Carbon-14 continually decays, but violent collisions in the atmosphere between high-speed particles create a replenishing supply. Since isotopes generally have the same chemical properties, despite their difference in mass, animals and plants incorporate carbon-14 into their cells and tissues. Plants, for example, use CO_2 to build carbohydrates, and animals eat these molecules. While the organism is alive, it contains about the same carbon-14 as the atmosphere. But when the organism dies, carbon-14 is no longer incorporated and slowly decays. This decay constitutes a clock that researchers can use to determine how long ago an organism died. Nitrogen-14 is a gas that leaks away, but researchers determine age by measuring the amount of carbon-14 and comparing it to the amount in the atmosphere when the organism was alive.

Although the technique works well, it has limitations. The amount of carbon-14 steadily decreases, and after about 50,000 years or so scientists have trouble measuring the little that remains. Another caveat is that the carbon-14 level in the atmosphere is not constant. Researchers have studied these variations (in many cases using techniques described in this chapter) and must take these values into account when calculating the age of a sample.

Pollen also provides researchers an indication of past climate. Seed plants release a sticky powder called pollen that contains grains consisting of protected reproductive cells. The plants reproduce by transferring these cells to other parts of the plant or to other plants. Sometimes the transfer takes place with the help of insects such as bees that unwittingly pick up the grains and move them from place to place in search of nectar. In other cases, pollen drifts with the wind. Pollen is abundant in certain times of the year, much to the dismay of allergy sufferers, who experience bouts of hay fever.

The durability of pollen means that it can last a long time, and since plants cannot aim the grains, they produce a large number of them in the hope that some will find the target. Scientists who study pollen grains and fossils can get some idea about which species of plants thrived and their relative abundance in past times. This information provides further clues about the climate in these periods.

Anne de Vernal and Claude Hillaire-Marcel of the Geochemistry and Geodynamics Research Center in Montreal, Canada, recently studied pollen grains recovered from marine sediments around Greenland. The researchers used pollen counts to reconstruct changes occurring in Greenland's vegetation, which would vary with the amount of available land—in other words, the amount of land not covered in ice. These variations reflect the responses in Greenland's ice sheet to climate changes in the past. Armed with this knowledge, scientists will be in a better position to predict how Greenland will respond to the present climate changes and any further variation that may occur in the future. These studies are critical because a melting of most of Greenland's ice would raise sea level about 23 feet (7 m).

The researchers discovered several periods in the last million years in which pollen suggests a substantial reduction in ice cover. Publishing their report in a 2008 issue of *Science,* de Vernal and Hillaire-Marcel note that their data shows "a long-term sensitivity of the Greenland ice sheet to warm temperatures. Among warm climate intervals of the last million years, MIS 11 [a marine isotope stage that refers to a period about 425,000 to 375,000 years ago] stands out in terms of forest vegetation spreading over southern Greenland." Greenland's ice sheets are susceptible to climate changes, although researchers must conduct additional studies to determine exactly how much ice melts.

MODERN TIMES—CONSULTING THE WRITTEN RECORDS

In addition to evidence provided by trees, fossils, and pollen, more recent information is available from another and even more reliable source—weather records. Thousands of weather stations, satellites, and observatories continuously monitor the weather. Huge volumes of data keep weather and climate researchers well informed about current conditions. But the recording of weather observations with scientific precision is a relatively new development, beginning in the middle of the 19th century.

Prior to the onset of precise observations, people had been making notes on the weather since the beginning of civilization. Ancient civilizations such as those of the Egyptians and the Sumerians monitored the weather, the seasons, and the rise of floodwaters from nearby rivers in order to maximize crop yields. Farmers throughout history have made such observations, keeping logs or almanacs for future reference. Other documents containing weather data include ships' logs, diaries and notes of travelers, and early newspapers and books. Although sketchy, these data can be mined for the prevailing conditions in historical times.

The systematic collation and assimilation of large amounts of weather data had to await the efforts of pioneers such as the American scientist Joseph Henry (1797–1878). Henry was the first secretary of the Smithsonian Institution, a research institute and museum founded in 1846. One of Henry's initial projects was the development of a weather reporting system. Henry set aside some money from the budget to enlist a network of weather observers to report conditions from all over the country. By 1849, he had the network up and running.

A great advance in Henry's system was the means by which observers made their reports to the Smithsonian. In 1837, the American inventor Samuel F. B. Morse (1791–1872) patented an electrical telegraph to send messages as coded signals via wires. By the 1840s, telegraph wires and communication stations were common. Henry employed this rapid means of communication so that the observers' reports were timely, passing along storm warnings and other important information. The network of observers grew, encompassing the United States from north to south, and included stations in Canada and Mexico. Outfitted with the necessary instruments, these early weather stations issued accurate weather bulletins. With this system in place, Henry became one of the founders of weather services and forecasting.

Weather services expanded over time, becoming more comprehensive. Historical climatologists can access much of the information compiled by these reporting systems, which serves a valuable climatic function (in addition to the storm warnings). For example, these records are critical in the calculation of the average temperature increases observed over the last century. The records also provide standards to refine paleoclimatology techniques—researchers can check the accuracy of the techniques by comparing their results with the recorded observations in the period in which these observations are available.

USING THE DATA: MODELING EARTH'S CLIMATE

Researchers have explored the past climate with paleoclimatology, probing ancient sea levels and the responses of Greenland's ice cover to climate changes, among other important topics. But scientists also want to put all of this data together and develop models of Earth's climate. Paleoclimatology becomes critical because knowledge of previous behavior is usually instrumental in predicting the future. Such models would be frameworks that researchers could use to test the effects of continued changes in climate, atmospheric composition, ocean currents, and so on.

A model is a representation of some object or process. In order to study and understand complicated systems, researchers often extract and isolate what they consider the essential components and their interactions. These variables go into the model so that its behavior mimics the real system as closely as possible. A model can be physical, such as that employed by engineers who build a miniature airplane or building and test its properties before they invest money in the real thing. Or a model can be a set of mathematical equations or a computer program or simulation. These paper or digital models show how the system's properties and behavior might change when the variables change. Important variables for climate models include temperature, atmospheric pressure and composition, solar energy input, water phases, and ocean levels and circulation.

Supercomputers are used in climate models as well as weather forecasting. These computers are located at the Office's headquarters in Exeter, Devon, in the United Kingdom. *(Michael Donne/Photo Researchers, Inc.)*

Models are only as good as the data that goes into them. To develop useful models, paleoclimatologists must be careful to exclude unreliable or questionable data. These models also rely on meteorologists, climatologists, physicists, chemists, and other scientists to provide reasonable theories by which the components of Earth's climate interact. And if researchers mistakenly omit an important variable, the model will almost certainly return misleading results.

Gavin A. Schmidt, a researcher at NASA's Goddard Institute for Space Studies (GISS) in New York City, summarized the fundamentals of climate modeling in a 2007 article, "The Physics of Climate Modeling," posted on GISS's Web site. "The task climate modelers have set for themselves is to take their knowledge of the local interactions of air masses, water, energy, and momentum and from that knowledge explain the climate system's large-scale features, variability, and response to external pressures, or 'forcings.' That is a formidable task, and though far from complete, the results so far have been surprisingly successful."

Climate modeling is not the same as weather forecasting. A weather forecast answers the question of what specifically will happen tomorrow or next week. Climate models offer a description of the general state and variability of conditions on the planet. These models attempt to answer questions such as how average temperatures or sea levels may rise if greenhouse gases continue to accumulate in the atmosphere.

As Schmidt noted, "More than a dozen facilities worldwide develop climate models, whose ability to simulate the current climate has improved measurably over the past 20 years. Interestingly, the average across all models almost invariably outperforms any single model, which shows that the errors in the simulations are surprisingly unbiased." When errors tend to cancel, this means that they straddle the actual value—some of the results are higher and some are lower. When averaged, the high values offset the low ones, so the outcome is more accurate than any one individual model.

Although climate models can presently address a host of issues, they are enormously complicated. Models are intended to simplify complex problems, but only so much simplification is permissible before the model becomes misleading.

Another problem is that models do not always agree, and sometimes the disagreement is with scientific measurements instead of another model—which means the model must be wrong. A press release posted on e! Science News on May 7, 2008, described a study of the Ohio State professor David Bromwich and his colleagues that compared Antarctic models with actual measurements. The result offered good news as well as bad news. "The models' predictions covering the last 50 years broadly follow the actual observed temperatures and snowfall for the southernmost continent, although the observations are very variable." Spoiling the good news to some extent was the bad news "that a similar comparison that includes the entire last century is a poor match. Projections of temperatures and snowfall ranged from 2.5 to five times what they actually were during that period." The models worked well when compared to the last 50 years but failed when the observations extended to 100 years.

The failure may be due to the inclusion of inaccurate amounts of water vapor in the models. Although the models incorporate water vapor, Bromwich noted in the news release that "we don't have anything to actually measure the amount of water vapor over the Antarctic continent." According to Bromwich, "Regarding water vapor over mainland Antarctica, the models just have to be wrong."

GLOBAL CLIMATE CHANGE

Researchers need more data, including proxy climate data from paleo-climatology, to improve on these results. The stakes are extremely high. Technology and industry have made life easier for people and greatly boosted economic productivity, but they have also increased the emission of greenhouse gases such as carbon dioxide. Potential effects of these increased emissions include accelerated warming, melting polar ice, and rising sea levels. But changing the technology and industrial activities that are responsible for these emissions will be costly and disruptive—many important processes, such as coal- and oil-based electric utilities and gasoline-fueled automobile engines, release greenhouse gases. Researchers need to specify the danger with clear and persuasive evidence in order to convince the world to take whatever action may be necessary.

Many scientists have turned to satellites for help. Orbiting instruments provide a global perspective that is especially beneficial when the object under study is the entire globe. The climate changes being studied are worldwide, and satellites that circle the globe in a few hours can provide imagery and instrument readings from a broad swath of Earth.

Recently, the Texas A & M University researcher Andrew Dessler and his colleagues used a NASA satellite called *Aqua* to measure *humidity* in the lower regions of the atmosphere. Humidity is a measure of the amount of water vapor in the air. Water vapor is a greenhouse gas, and it is the most abundant such gas in Earth's atmosphere. Dessler and his colleagues used the water vapor data in combination with CO_2 data and temperature measurements to study the interactions between these factors.

In a NASA news release issued on November 17, 2008, Dessler said, "Everyone agrees that if you add carbon dioxide to the atmosphere, then warming will result. So the real question is, how much warming?" The researchers investigated this question by measuring water vapor. Climate models indicate that water vapor is part of a positive feedback loop—higher temperatures cause greater evaporation and more water vapor, which as a greenhouse gas leads to higher temperature, and so on. "The difference in an atmosphere with a strong water vapor feedback and one with a weak feedback is enormous," observed Dessler. The finding of this research team indicates if Earth's average temperature rises another 1.8°F (1.0°C), the resulting increase in water vapor will cause the planet to hold onto another two watts of power per 10.8 ft^2 (1 m^2).

Two watts is not very much—a typical lightbulb uses 60 watts. But this effect will take place worldwide, which means the overall amount is huge. According to the news release, "Because the new precise observations agree with existing assessments of water vapor's impact, researchers are more confident than ever in model predictions that Earth's leading greenhouse gas will contribute to a temperature rise of a few degrees by the end of the century."

Models and data have given the IPCC enough confidence to attribute the recent warming trend to human activity. In the *Climate Change 2007: Synthesis Report,* issued in November 2007, the IPCC scientists noted, "Global GHG [greenhouse gas] emissions due to human activities have grown since pre-industrial times, with an increase of 70% between 1970 and 2004." The scientists concluded that there is a link between these gases and the rise in temperature. "Most of the observed increase in global average temperatures since the mid-20th century is *very likely* due to the observed increase in anthropogenic GHG concentrations." IPCC defines the term very likely to mean a greater than 90 percent probability.

CONCLUSION

More data and observations will be needed to lend further support for climate models. Model accuracy has been increasing, as judged by their conformance to past climate and paleoclimatological data, but refinements could turn "very likely" into "almost assuredly." Ironclad evidence persuades more skeptics.

In addition to the recent warming trend, paleoclimatology and its associated climate models can shed light on other critical phenomena. One of the most important functions of the atmosphere is to protect life on the surface from harmful radiation. Ozone in the atmosphere, for example, absorbs ultraviolet radiation from the Sun, reducing the amount of this dangerous radiation that reaches the surface. But the level of atmospheric ozone has varied in the past—and in 1985, the British scientists Joseph Farman, Brian Gardiner, and Jonathan Shanklin reported significant reductions in ozone over Antarctica—a "hole" in the protective ozone layer.

Ozone is a molecule made of three oxygen atoms (O_3). Although often a component of smog, most of the ozone in the atmosphere resides in the stratosphere, 16 to 80 miles (10–50 km) above the surface. The Sun's radiation ionizes oxygen, and the ions combine and produce a layer of ozone that blocks much of the biologically hazardous ultraviolet portion of the spectrum.

Certain chemicals, however, react strongly with ozone and decrease the number of ozone molecules. One of these chemicals is chlorine, an element that forms part of compounds such as a class of substances known as chlorofluorocarbons (CFC), which in the past were commonly used in cleaning solvents and refrigerants (fluids used in cooling systems). When CFCs reach the upper layers of the atmosphere, solar radiation disrupts the chemical bonds, releasing chlorine. The thinned areas of ozone resulted in significant increases in ultraviolet radiation at the surface. Elevated exposure to this radiation increases health risks such as skin cancer.

Alarmed, officials from many countries agreed to an international treaty known as the Montreal Protocol on Substances That Deplete the Ozone Layer. Based on discussions at a meeting held in Montreal in 1987, the Montreal Protocol called for a reduction in the use of ozone-reducing compounds. The treaty became effective in 1989. Most nations eventually signed, and CFC production was phased out in 1996. Since then, ozone levels have begun to recover.

A new study shows that ozone recovery may not be uniform—some areas may experience lower levels of ozone than others. Feng Li, a researcher at the Goddard Earth Sciences and Technology Center in Baltimore, Maryland, and his colleagues examined an extensive model that included atmospheric chemistry, wind patterns, and solar radiation. The scientists discovered that greenhouse gases affect winds that in turn affect the distribution of ozone in the atmosphere. As a result, rising concentrations of greenhouse gases will disturb ongoing ozone recovery in certain regions. A NASA press release issued on April 9, 2009, announced, "In Earth's middle latitudes . . . ozone is likely to 'over-recover,' growing to concentrations higher than they were before the mass production of CFCs. In the tropics, stratospheric circulation changes could prevent the ozone layer from fully recovering."

Climate models will continue to be essential tools in the study of Earth's changing conditions. And paleoclimatology will continue to be vital in the construction and testing of these models. As the journalist Richard Kerr wrote in a 2006 issue of *Science,* "Researchers worry that if they cannot recall the distant climatic past, the world may be condemned to repeat it." The Montreal Protocol is a successful example of international cooperation in the face of adverse changes. Recent trends such as global warming and increasing amounts of greenhouse gases may also foreshadow considerable trouble, but world leaders need to learn more about this trouble and its potential effects in order to take appropriate action.

CHRONOLOGY

1788	The Scottish geologist James Hutton (1726–97) presents his views that Earth is much older than suggested by the then-prevalent unscientific beliefs, which held that Earth was only a few thousand years old.
1815	The Swiss mountaineer Jean-Pierre Perraudin (1767–1858) notes that scratches in the ground and the position of boulders in certain valleys suggest glaciers once extended much farther than they do at present.
1837	Based on geological evidence, the Swiss-American scientist Louis Agassiz (1807–73) proposes that Earth experienced an ice age in its past.
1849	The American scientist Joseph Henry (1797–1878) develops one of the first telegraphic weather services.
1890s	The American scientist Andrew Ellicott Douglass (1867–1962) develops the field of dendrochronology.
1895	The Swedish chemist Svante Arrhenius (1859–1927) proposes that atmospheric CO_2 could influence Earth's temperature.

1913	The French scientists Maurice Fabry (1867–1945) and Henri Buisson (1873–1944) discover the ozone layer.
1949	The American chemist Willard F. Libby (1908–80) and his colleagues develop the technique of radiocarbon dating.
1970	The U.S. government establishes NOAA.
1980s	Rising temperatures and melting glaciers begin to alarm scientists.
1985	The British scientists Joseph Farman, Brian Gardiner, and Jonathan Shanklin discover significant reductions in ozone over Antarctica.
1988	The Office of Polar Programs at the National Science Foundation, a major U.S. agency that funds scientific research, initiates the Greenland Ice Sheet Project 2 in order to obtain extensive ice cores for paleoclimatology.
1989	The Montreal Protocol on Substances That Deplete the Ozone Layer takes effect.
1993	The Greenland Ice Sheet Project 2 reaches bedrock. Drillers retrieve an ice core 10,000 feet (3,053 m) deep.
2002	NASA launches *Aqua,* an Earth-observing satellite primarily intended to collect data on water and the water cycle.
2007	The IPCC releases *Climate Change 2007.*
2008	The Ohio State professor David Bromwich and his colleagues compare Antarctic models with actual measurements and find a good fit for the last 50 years but a poorer performance when the time period is extended to 100 years.

FURTHER RESOURCES
Print and Internet

Ahn, Jinho, and Edward J. Brook. "Atmospheric CO_2 and Climate on Millennial Time Scales During the Last Glacial Period." *Science* 322 (10/3/08): 83–85. The researchers studied gases trapped in Antarctic ice cores dating from 20,000 to 90,000 years ago.

Alley, Richard B. *The Two-Mile Time Machine: Ice Cores, Abrupt Climate Change, and Our Future.* Princeton, N.J.: Princeton University Press, 2002. The climate researcher Richard Alley discusses ice cores and how scientists use them to study the climate of the past and what may happen in the future.

De Vernal, Anne, and Claude Hillaire-Marcel. "Natural Variability of Greenland Climate, Vegetation, and Ice Volume during the Past Million Years." *Science* 320 (6/20/08): 1,622–1,625. The researchers studied pollen preserved in marine sediments to correlate climate changes, vegetation, and ice volume in Greenland.

e! Science News. "Global Climate Models Both Agree and Disagree with Actual Antarctic Data." News release (5/7/08). Available online. URL: http://esciencenews.com/articles/2008/05/07/global.climate. models.both.agree.and.disagree.with.actual.antarctic.data. Accessed July 1, 2009. Research comparing Antarctic models with snowfall and temperature data shows good agreement in the last 50 years but significant differences when the comparison includes the past 100 years.

Intergovernmental Panel on Climate Change. *Climate Change 2007: Synthesis Report.* Available online. URL: http://www.ipcc.ch/pdf/ assessment-report/ar4/syr/ar4_syr.pdf. Accessed July 1, 2009. The IPCC scientists review climate data and models.

Kerr, Richard A. "Looking Way Back for the World's Climate Future." *Science* 312 (6/9/06): 1,456–1,457. Kerr describes some recent findings in paleoclimatology.

Maasch, Kirk A. "The Big Chill." Available online. URL: http://www. pbs.org/wgbh/nova/ice/chill.html. Accessed July 1, 2009. Maasch describes the ice ages and how researchers study them.

Macdougall, Doug. *Frozen Earth: The Once and Future Story of Ice Ages.* Berkeley: University of California Press, 2006. Macdougall, a researcher at the Scripps Institution of Oceanography, tells the story of Earth's periodic ice ages.

National Academies. "'High Confidence' That Planet Is Warmest in 400 Years; Less Confidence in Temperature Reconstructions Prior to 1600." News release (6/22/06). Available online. URL: http://www8. nationalacademies.org/onpinews/newsitem.aspx?RecordID=11676. Accessed July 1, 2009. The NRC reports that Earth's temperature is the highest it has been in the last four centuries.

National Aeronautics and Space Administration. "Water Vapor Confirmed as Major Player in Climate Change." News release (11/17/08). Available online. URL: http://www.nasa.gov/topics/earth/features/ vapor_warming.html. Accessed July 1, 2009. Researchers use a NASA satellite to measure water vapor in the atmosphere.

———. "Climate Change and Atmospheric Circulation Will Make for Uneven Ozone Recovery." News release (4/9/09). Available online. URL: http://www.nasa.gov/topics/earth/features/ozone_recovery. html. Accessed July 1, 2009. A climate model indicates increasing levels of greenhouse gases will result in changes in ozone distribution.

National Oceanic and Atmospheric Administration. "Introduction to Paleoclimatology." Available online. URL: http://www.ncdc.noaa. gov/paleo/primer_care.html. Accessed July 1, 2009. NOAA's primer on paleoclimatology discusses how researchers learn more about Earth's past climate.

National Science Foundation. "Gas from the Past Gives Scientists New Insights into Climate and the Oceans." News release (10/3/08). Available online. URL: http://www.nsf.gov/news/news_summ. jsp?cntn_id=112395. Accessed July 1, 2009. This press release summarizes the findings of Jinho Ahn and Edward J. Brook that they published in their 2008 *Science* paper.

Riebeek, Holli. "Paleoclimatology: The Ice Core Record." Available online. URL: http://earthobservatory.nasa.gov/Features/Paleoclima tology_IceCores/. Accessed July 1, 2009. This short and informative article describes what scientists can learn by studying ice cores.

Schmidt, Gavin A. "The Physics of Climate Modeling." Available online. URL: http://www.giss.nasa.gov/research/briefs/schmidt_04/. Accessed July 1, 2009. Schmidt, a researcher at the GISS, reviews climate model mechanics.

Weart, Spencer R. *The Discovery of Global Warming: Revised and Expanded Edition.* Cambridge: Harvard University Press, 2008. Weart recounts the history of the recent discovery that Earth's climate is changing. The book describes science's typical zigzag path toward the truth.

Web Sites

Earth Observatory: Paleoclimatology. Available online. URL: http://earthobservatory.nasa.gov/Study/Paleoclimatology/. Accessed July 1, 2009. NASA's Earth Observatory hosts a series of articles on paleoclimatology. Topics include an introduction to the subject, marine sediments, ice cores, geological clues, historical records, and climate models.

National Oceanic and Atmospheric Administration. Available online. URL: http://www.noaa.gov/. Accessed July 1, 2009. NOAA's Web site contains news and information on this large agency and its research and management efforts.

NOAA Paleoclimatology. Available online. URL: http://www.ncdc.noaa.gov/paleo/paleo.html. Accessed July 1, 2009. The Web site of this branch of NOAA's National Climatic Data Center includes data from dendroclimatology, ice cores, sediments, coral reefs, pollen, and historical records.

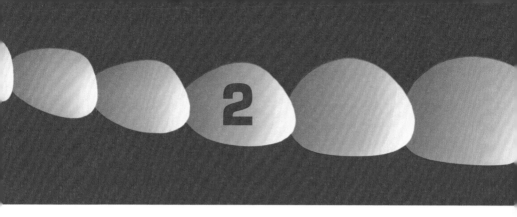

Polar Climate—
A Sensitive
Indicator of Change

Earth rotates on its axis, which is tilted about 23.5 degrees—instead of being perpendicular to the plane in which the planet orbits, the axis makes an angle of 23.5 degrees with the perpendicular (which is an imaginary line drawn straight up from the orbital plane, making a right angle with it). Because of this tilt, the Northern and Southern Hemispheres receive a varying amount of sunlight during the course of year, as Earth revolves around the Sun. This is the primary cause of seasonal variations in temperature. Earth's axis also defines two special points where the axis meets the surface—one on top and one on the bottom. These points are known as the North Pole and the South Pole. Areas surrounding a pole are called polar regions or polar environments. The area surrounding the North Pole is called the Arctic (from the Greek *arktikos,* meaning northern), and the region of the South Pole is called the Antarctic region (from *antarktikos,* meaning the opposite of north).

The climate at the poles is harsh and frigid. Polar regions receive little solar radiation because they are on and around the axis instead of facing the Sun. As a result, the Sun does not rise high in the sky, and fails to deliver much warmth. (See the figure on page 61.) Polar regions experience year-round snow and ice, although the amount varies with the seasons.

Researchers have widely documented a number of climate changes recently of great concern to scientists and others. Many of these climate changes are global in scope but not uniform—the changes in some regions have been greater or lesser than the average. Polar climates seem to be more sensitive than other regions, especially the Arctic, which has seen higher than average departures from normal conditions. For example, winter temperatures in parts of the Arctic have risen more than four times higher than the global average value over the last 100 years (the global average is 1.33°F [0.74°C], according to the Intergovernmental Panel on Climate Change [IPCC]), and Arctic *sea ice*—frozen ocean water—hit a record low for recent times in 2007. General changes in climate may be getting amplified in the extreme polar environments.

Although both polar regions have extreme environments, they are quite different and have seen different changes in recent years. While sea ice is disappearing in the Arctic, it is increasing in extent in the Antarctic. Researchers who study polar climates want to understand how these climates work, how they affect the region's wildlife, and the nature of climate changes that have been occurring. Climate change in the sensitive polar regions is particularly important to monitor and study since it might be the harbinger of what is to come for the rest of the globe. The goal of much of the research in these icy frontiers of science is to document recent changes and to explore their causes and effects. The goal of this chapter is to describe the progress researchers have made and what remains to be done.

INTRODUCTION

In the winter, low temperatures in the Arctic can reach -40°F (-40°C) or lower. Winter temperatures in the Antarctic usually vary from -40°F (-40°C) to -90°F (-68°C). The lowest recorded temperature in the world is -129°F (-89°C), measured at Russia's Vostok Station in the Antarctic on July 21, 1983. (Since this region is in the Southern Hemisphere, July is a winter month, which is the opposite of the Northern Hemisphere.) Summer temperatures are tens of degrees higher in general, but in many parts of the poles, the snow cover reflects a great deal of sunlight instead of absorbing it, reducing the radiation's ability to raise temperatures.

The North Pole is located at a snow- and ice-covered point in the Arctic Ocean. Most people consider the Arctic region to include the area inside the Arctic Circle, as shown in the following figure. This area includes the northern parts of Canada, Alaska, Greenland, Norway, Sweden, Finland, Russia, and Iceland. The Arctic Circle is the latitude above

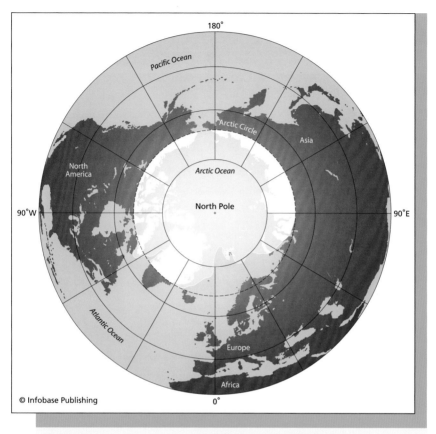

The Northern Hemisphere, showing the Arctic and its ice cover

which the Sun does not set on the *summer solstice*—the day, usually June 21, at which the Northern Hemisphere is most inclined to the Sun, so it is the longest day in this hemisphere (and lasts all 24 hours in the Arctic). In the *winter solstice,* the Sun does not rise in the region north of the Arctic Circle. And at the North Pole and its vicinity, the Sun does not set from late March to late September—six months of sunlight—and does not rise in the other months, so there are about six consecutive months of darkness during the year. Lower latitudes within the Arctic Circle also experience long periods of light and darkness, though not as extreme.

Antarctica, the fifth largest continent, contains the South Pole, as illustrated in the figure below. The Antarctic Circle is the southern version of the Arctic Circle and defined similarly. Antarctic territories dif-

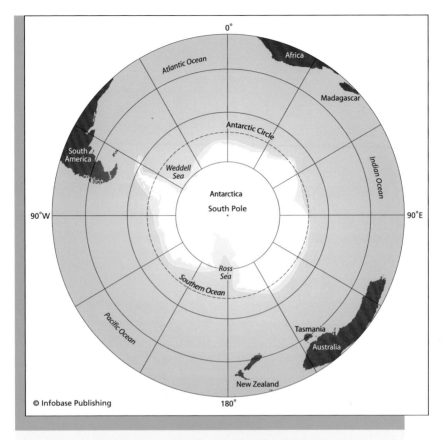

The Southern Hemisphere, showing Antarctica and its ice cover

fer from their northern counterparts in that the Antarctic contains a large landmass instead of being mostly ocean. But both poles are capped with snow and ice. Some of the land in the lower latitudes of the Arctic consists of *tundra*—cold plains with some vegetation (though not trees, which do not thrive in the cold climate)—but Antarctica is snow, ice, and the occasional rocky outcrop.

Parts of the Arctic region are habitable. Native inhabitants of isolated northern sections of Greenland and Canada are hardy people called the Inuit, who survive mostly on hunting and fishing. (Inuit is plural, meaning "the people." Inuk refers to a single person.) Antarctica has no native inhabitants.

Frigid and inaccessible, the North and South Poles have long enticed explorers to attempt the challenge of reaching them. The American explorer Robert Peary (1856–1920), along with Matthew Henson and several Inuit, reported reaching the North Pole on April 6, 1909, but this claim was not verified and remains controversial to this day. The British explorer Sir Walter "Wally" Herbert (1934–2007) led the first expedition that indisputably set foot on the North Pole on April 6, 1969. British, French, and American naval expeditions in the late 1830s established by 1840 that Antarctica was a continent, and the Norwegian explorer Roald Amundsen (1872–1928) led a land expedition that reached the South Pole on December 14, 1911.

Traveling across the desolate and treacherous ice around the poles is extremely difficult. Navigating and keeping track of location is not easy. The position of the Sun or the stars can be a guide, when they are visible. For instance, because of Earth's rotation, the stars sweep across the sky. In the Northern Hemisphere, stars appear to circle around a point above the North Pole. Polaris, the North Star, is located near this point and does not move very much, and other stars revolve around it. (Viewers in the Northern Hemisphere who live in the Tropics or the equatorial regions only observe a small arc of this motion for most stars in their night sky.) When an explorer is standing on the North Pole, he or she is on the axis of rotation, and Polaris appears directly overhead.

The normally reliable compass is not much use at the poles. Compasses detect Earth's magnetic field, which behaves similarly to a bar magnet with two poles. But the magnetic poles do not coincide with the geographical poles; the north magnetic pole is currently about 500 miles (800 km) from the geographic North Pole, and the south magnetic pole is about 1,750 miles (2,800 km) from the geographic South Pole. The discrepancy is mostly noticeable in the polar regions. At the North Pole, a compass would point toward the south, while at the north magnetic pole a compass would point straight down! These days, researchers who venture to the poles or who reside in polar research stations have sophisticated tools such as global positioning systems (GPS) that employ satellites and receivers for pinpoint accuracy.

RECENT CLIMATE CHANGES

In the last decade, climatologists have found disconcerting changes occurring in the Arctic. Mark C. Serreze, a researcher at the National Snow and Ice Data Center (NSIDC) in Boulder, Colorado, recently reviewed these changes in the winter 2008/2009 issue of *Witness the Arctic,* a publication

National Snow and Ice Data Center

Scientific research often generates a huge amount of data. Weather and climate data include charts, maps, and statistics consisting of many different kinds of measurements, such as regional precipitation and temperature. Researchers publish a small amount of this data, usually summarized in the form of tables or graphs, in science journals. But other researchers who wish to build upon previous studies need access to much more of the data so that they can compare and contrast the values to those of their own measurements. To achieve this general purpose, data "warehouses" collect and store certain kinds of data and make this information available to researchers all over the world. Data management and distribution is one of the main goals of the National Snow and Ice Data Center, a branch of the Cooperative Institute for Research in Environmental Sciences at the University of Colorado.

NSIDC got its start back in the analog days before data began to be digitized and stored on computers. In the 1957–58 International Geophysical Year, which consisted of a coordinated effort to conduct a series of important observations—particularly in the Arctic and Antarctic regions—the World Data Center for Glaciology was set up to archive this data. Originally under the American Geographical Society, the World Data Center for Glaciology transferred to the United States Geological Survey (USGS) in 1970, then went to the University of Colorado in 1976, where it found its current home at the Cooperative Institute for Research in Environmental Sciences and its current name in 1982.

In addition to other duties, scientists at NSIDC also conduct their own research. With funds from the National Aeronautics and Space Administration (NASA), the National Oceanic and Atmospheric Administration (NOAA), and the National Science Foundation (NSF), researchers study many different aspects of the cryosphere, including climate variability at the polar regions and its impact on inhabitants and environment.

of the Arctic Research Consortium of the United States. Serreze wrote, "Recognition that the Arctic was in the midst of widespread change began to emerge in the mid and late 1990s. From work by Bill Chapman and John Walsh of the University of Illinois and Jim Hurrell at the National Center for Atmospheric Research (NCAR), it became clear that northern Eurasia and North America had experienced substantial warming since about 1970, largest during winter and spring and partly compensated by cooling over northeastern North America."

Other significant changes accompanied the warming, affecting glaciers, vegetation, and *permafrost*. "Mark Dyurgerov and Mark Meier of the University of Colorado showed that the mass balance of arctic glaciers had become persistently negative, paralleling a global tendency. Other studies were finding increased plant growth in the Arctic, northward advance of the treeline, and an increased frequency of forest fires. Vladimir Romanovsky and Tom Osterkamp of the University of Alaska Fairbanks found that Alaskan permafrost was warming and locally thawing."

The NSIDC plays an important role in tracking the changes in the Arctic and Antarctic regions. As discussed in the sidebar on page 34, NSIDC's specialty is the *cryosphere*—the regions containing frozen water, such as sea ice, ice sheets and caps, glaciers, snow, and permafrost. (Cryosphere derives from a Greek word *kryos* meaning "cold.")

Climatologists had long suspected that the Arctic would be extremely susceptible to change. An important factor in the Arctic's sensitivity is *albedo*—the fraction of light an object reflects, which is a measure of its surface reflectivity. In an essay published in NOAA's Arctic Theme Page, Serreze wrote, "That the Arctic should be especially sensitive to climate change was recognized in the 19th century. The primary reason for this sensitivity is that an initial warming (or cooling) sets in motion a chain of events that amplify the warming or cooling. This chain of events is known as the albedo feedback." Snow is highly reflective (so much so that it can result in snow blindness in unprotected explorers), so a lot of radiation that strikes the polar regions is reflected rather than absorbed. If a warming trend starts to melt the snow and exposes less reflective ground or rock, this means that more radiation gets absorbed—and this energy warms the region even further, resulting in more melting, and so on. In this manner, an initial change gets amplified. The same holds true for the Antarctic, although the effect is not as

Arctic sea ice, as measured by a NASA satellite *(Science Source)*

great because most of the landmass is around the pole, where it stays colder, rather than at the periphery, as it is in the Arctic.

In summer 2007, sea ice in the Arctic reached its lowest extent since satellite observations began in the late 1970s, and although sea ice in the Antarctic is not diminishing, other snow and ice structures are melting in both polar regions. For example, scientists have observed the splintering of a bridge of ice connecting the Wilkins Ice Shelf, a large region of floating ice near the Antarctic Peninsula, to Charcot and Latady Islands. Wilkins Ice Shelf began to melt and recede in the 1990s, and the changes have been rapid. In a press release posted on April 4, 2009, at Science*Daily*, Angelika Humbert of the Institute of Geophysics at Münster University in Germany said, "During the last year the ice shelf has lost about 1,800 sq km [690 miles^2] or about 14% of its size. The break-up events in February and May 2008 happened in just hours, leaving the remaining part of the ice bridge in a fragile situation."

Global warming can certainly be expected to contribute to the retreat of snow cover and glaciers, but it may not be the only factor. David Holland, a researcher at New York University; Malte Thoma, a researcher at the Alfred Wegener Institute for Polar and Marine Research in Germany; and their colleagues found that in certain cases the melting of glaciers in Greenland and the Antarctic have a cause other than global warming. The journalist Richard A. Kerr noted in a 2008 issue of *Science* that the "new studies point to random, wind-induced circulation changes in the ocean—not global warming—as the dominant cause of the recent ice losses through those glaciers." The changes result in warmer water bathing the coasts, which heats up the dwindling ice sheets.

These new studies do not reduce the alarm over climate change—the circulation patterns are part of the change, and the glaciers continue to melt, no matter what kind of effect is directly responsible. But this research points out that scientists still have some work to do in order to understand polar climates—and figure out what is causing the changes, the impact these changes may have, and what can be expected in the future.

ARCTIC RESEARCH

Polar climate research is a frontier of science because of its cold, forbidding environment as well as its scientific mysteries. The inhospitable conditions hamper a great deal of research effort in this area. Remote sensing can substitute for some of the tasks, but jobs such as obtaining samples and making direct measurements require the presence of researchers—and it usually is a cold and uncomfortable job.

A researcher measuring the density of snow at a weather station at Tarfala, Sweden *(David Hay Jones/Photo Researchers, Inc.)*

Many weather or research stations in the Arctic are situated on small islands such as Axel Heiberg Island, which is home to the McGill Arctic Research Station (latitude 79°26' N, longitude 90°46' W), and the Svalbard archipelago, which houses several stations including the Arctic Research Station (latitude 78°56' N, longitude 12° E) of the United

Kingdom's Natural Environmental Research Council. Other stations and observatories dot Greenland and the northern regions of Russia, Scandinavia, Canada, and Alaska.

As a spark for further research, the World Meteorological Organization (WMO) and the International Council for Science organized the International Polar Year, which ran from March 2007 to March 2009. There have been three other polar years—1882–83, 1932–33, and 1957–58—which sponsored a great deal of polar research. The most recent International Polar Year ran two years so that researchers could have sufficient time for both the Arctic and Antarctic. Two hundred projects involving scientists from many different countries studied climate as well as polar biology and geology.

Staying at or near the North Pole is hazardous because there is no terra firma beneath one's feet. Although sea ice covers the Arctic Ocean at and around the North Pole, it is usually only around six to 12 feet (1.8–3.6 m) thick and getting thinner due to the changing climate. Shifting ice and cracks are serious threats. Researchers tread carefully and do not stay long.

Beginning in 2000, the NSF, a U.S. agency that funds scientific research, sponsored the North Pole Environmental Observatory. Each spring, researchers traveled to the pole to make field measurements and to set up unmanned observatories that recorded data throughout the rest of the year. Measurements include temperature, ice thickness, and the properties of the Arctic Ocean water. The program ran through 2008.

But Arctic environments are not always covered in snow and ice. The lower latitudes enjoy temperatures warm enough to expose the ground, at least for some portion of the year, although much of it is permafrost. Researchers have not neglected these regions.

For example, the University of Alaska Fairbanks researcher Chien-Lu Ping and his colleagues recently took samples of permafrost from more than 100 locations in Alaska. Ping and his colleagues dug deeper than previous researchers and sampled the frozen soil at depths of more than 3.3 feet (1 m). Chemical analysis of these samples revealed a previously unknown layer of organic material—material containing much carbon. This material comes from biological organisms and settles on the surface of the soil until it gets buried by the cycles of freezing and thawing experienced in this part of the world.

The storage of carbon-containing material in Arctic soil is important for several reasons. One of the most significant factors is its potential to release carbon dioxide, methane (a hydrocarbon containing one carbon atom and four hydrogen atoms), and other greenhouse gases. These gases, as described in chapter 1, trap infrared radiation and retain heat, creating a warming effect similar to what happens in a greenhouse. Many industrial processes and machinery, including internal combustion engines in automobiles, emit greenhouse gases; a rise in these gases correlates with the recent global warming trend and according to the IPCC is a likely contributor. Another source of these gases is chemical processes that decompose organic material. Researchers worry that rising temperature could accelerate these processes, resulting in the release of more greenhouse gases, leading to higher temperatures, and so on in a positive feedback loop.

CARBON IN THE ARCTIC OCEAN

Concern about carbon is not limited to the soil—the Arctic Ocean and its seafloor are concerns as well. In addition to organic matter from marine organisms, oceans are the final destination of a great deal of organic matter from land surfaces. Terrestrial plants and animals live and die, leaving a considerable amount of biological material for rain to sweep into rivers. Rivers flow into seas, carrying much of this organic matter into the shallow portion of the ocean near land—the continental shelf. Organic matter accumulates on the ocean floors, usually stacked in sediments. A lot of Earth's carbon is stored in this manner. Ocean basins act as carbon sinks, taking in a great deal of carbon. As such, they reduce the amount of greenhouse gases released into the atmosphere.

But temperature changes may alter this situation, especially in the Arctic, where temperatures are rising more rapidly than elsewhere. The Arctic Ocean does not receive input from, say, a river like the Mississippi, which drains a huge swathe of fertile land in the United States, but a large amount of material from bogs and similar environments ends up in the Arctic Ocean. Scientists are concerned that rising temperatures could vanquish sea ice, churn up sediments, and thaw permafrost buried under the ocean floor, releasing significant quantities of greenhouse gases such as carbon dioxide and methane.

Researchers have already found some evidence to support this idea. Igor Semiletov, a researcher at the International Arctic Research Center in Fairbanks, Alaska, and his colleagues measured the amount of methane dissolved in water in the East Siberian Arctic Shelf, part of the continental shelf of northern Russia. This large shelf extends more than 900 miles (1,440 km) into the Arctic Ocean. Trapped in its sediments is a great deal of carbon, which had appeared to be safely buried. But the team led by Semiletov found that the levels of methane are rising. In a press release issued on December 15, 2008, by the International Arctic Research Center, Semiletov noted, "The concentrations of the methane were the highest ever measured in the summertime in the Arctic Ocean. We have found methane bubble clouds above the gas-charged sediment and above the chimneys going through the sediment."

Methane is a much stronger greenhouse gas than carbon dioxide (though less prevalent in the atmosphere), so Semiletov and his team's finding is alarming. Arctic researchers will continue to monitor levels of carbon and greenhouse gases in land and water.

SEA ICE

Other active areas of research involve the dwindling Arctic sea ice. Ponds and lakes freeze when the temperature dips below the freezing point of water, and the same is true for the ocean, except the salt in seawater affects the freezing point—freshwater freezes at 32°F (0°C) at normal atmospheric pressure, but seawater tends to freeze at a slightly lower temperature, about -28.8°F (-1.8°C). Water expands as it becomes solid, so ice is less dense than liquid, which means ice floats in water. The frigid conditions in polar regions result in a blanket of ice covering the surface. This sea ice covers about 7 percent of the total area of the world's oceans. Its high albedo raises the overall planet's reflectivity, bouncing a great deal of solar radiation back into space.

During the winter, sea ice accumulates a layer of snow. This snow melts in the summer, collecting in pools. Sunlight warms this water more quickly than ice because water absorbs more solar radiation. The pools of water may drain through cracks in the ice or melt all the way through, forming a thaw hole. Some of the ice melts as well, but some of it endures through the summer months.

Scientists monitor the extent and thickness of sea ice as a gauge of polar climate changes. Satellite images show the extent of ice, and researchers can sample the thickness at various points by drilling. Sam-

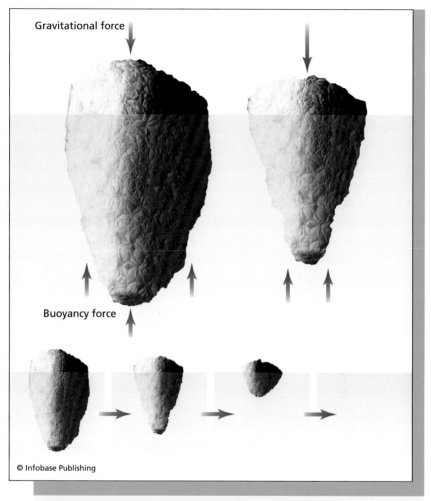

Gravitational force

Buoyancy force

© Infobase Publishing

The top of the figure shows the force of gravity balanced by buoyancy—an upward force equal to the weight of the water displaced by the floating object. The panels at the bottom of the figure illustrate that the melting of floating ice does not alter the water level.

pling introduces uncertainty (as well as being time-consuming and dangerous), so researchers have turned to satellites to estimate the thickness as well as the extent of sea ice. The method involves an instrument called an altimeter, which works like radar except its purpose is to measure the satellite's altitude—the distance from the satellite to the surface. Altimeters perform this measurement by bouncing a pulse of electromagnetic radiation off the surface; multiplying the travel time by the speed of light gives the distance traveled. Researchers can map Earth's surface elevation precisely with satellite altimeters, and by comparing the height of sea ice with average sea levels, polar climate scientists can estimate the thickness of sea ice.

Scientists at the Centre for Polar Observation and Modelling (CPOM), part of the National Centre for Earth Observation, recently used *Envisat*, a satellite of the European Space Agency (ESA), to estimate Arctic sea ice thickness. The researchers, led by Katherine Giles, found that Arctic sea ice thinned by an average of 10 percent in the winter of 2007–08 when compared to the last five winters, which were relatively stable. Large areas of the western Arctic sea ice lost about 19 percent of its thickness.

How will the reduction in Arctic sea ice affect weather and climate? Scientists are unsure, and this is an important issue to address. The melting of sea ice does not cause the sea level to rise—floating ice displaces an equivalent volume of water, so the change in phase from ice to water has no effect on sea level, as illustrated in the figure on page 41—but it is a harbinger of changes to come. In some respects, these changes may have advantages, such as allowing ships to sail through passages they could not otherwise navigate. But other consequences may not be so beneficial.

Science*Daily* recently reported a study of Arctic weather led by Erik Kolstad of the Bjerknes Centre for Climate Research in Bergen, Norway. Kolstad and his colleagues flew into storms and severe weather in order to make direct measurements of pressure, temperature, wind, and so on. This data will help researchers analyze weather patterns that might be changed with the reduction of sea ice. The Science*Daily* news release, issued on February 5, 2009, noted that "regions that have been covered by sea ice until now will be exposed to new kinds of severe weather." This is particularly true since Arctic weather systems often begin when cold air over the ice sheets moves to warmer water, resulting in the development of storms.

ANTARCTIC OBSERVATORIES AND RESEARCH

Unlike the Arctic, the Antarctic has a huge landmass—the continent of Antarctica—sitting at the pole. This gives the Antarctic a distinct ge-

ANDRILL facility in Antarctica *(Science Source)*

ography as well as distinct weather and climate conditions. Antarctica also gives scientists a lot of terra firma on which to establish research stations.

Naval expeditions in the 19th century established that Antarctica is a continent, and further exploration showed that it harbored no native inhabitants. (Antarctica is the only continent lacking a native population.) The same harsh conditions that prevented earlier settlement also blocked ideas of colonization in the 19th and 20th centuries, but researchers quickly moved in and set up stations. The Scottish scientist William S. Bruce (1867–1921) led an expedition in 1903 that resulted in the establishment of the first permanent base in the Antarctic at Laurie Island. Argentina has maintained a presence at this base, called Orcadas, since 1904.

Today the activities in the Antarctic abide by a set of rules established in the Antarctic Treaty, signed in 1959 by 12 nations (Argentina, Australia, Belgium, Chile, France, Japan, New Zealand, Norway, South Africa, Union of Soviet Socialist Republics, United Kingdom, and United States) with an interest in the region. The treaty went into effect in 1961, and a total of 47 countries have now signed it. These countries

McMurdo Station

McMurdo Station began in 1955 as a collection of tents known as Naval Air Facility McMurdo at which the U.S. Navy landed supply aircraft. Located on a barren outcrop of volcanic rock on Hut Point Peninsula on Ross Island, off the coast of Antarctica, the station takes its name from nearby McMurdo Sound, which was named for a British naval officer Archibald McMurdo, who mapped the area in 1841. McMurdo Station shares Ross Island (named after the British explorer James Clark Ross, who commanded the expedition in which McMurdo participated) with penguins and skuas.

About 85 buildings make up McMurdo Station, which can house more than 1,200 people, making it the largest base in the Antarctic. Icebreakers keep the harbor open and navigable to ships, and the base also has landing strips on the ice for airplanes. The station is like a small city, with dormitories, a power plant, firefighting equipment, shopping centers, a water distillation facility, and other structures, including scientific laboratories. Telephone and power lines link the buildings, along with water and sewer pipes. The station is generally free of snow in the summer—temperatures average a not exactly balmy 27°F (-3°C) in January (which is one of the warmer months in the Southern Hemisphere)—but gets a lot of snow in the middle of the year, with average temperatures falling to -18°F (-28°C) in August.

McMurdo Station serves as the main gateway to Antarctica for American researchers. For instance, personnel headed to other Antarctic research stations, such as the Amundsen-Scott South Pole Station, usually stop at McMurdo on the way. It is the last chance for these polar scientists to enjoy "deluxe" accommodations before they arrive at their smaller, more cramped destinations.

pledge to maintain free access for scientific research. Military activity is banned.

More than two dozen countries have established research stations in the Antarctic. The United States has several, including one, the Amundsen-Scott South Pole Station, right at the South Pole. (This station is named for the first person to reach the South Pole, the Norwegian explorer Roald Amundsen (1872–c. 1928), and for the British explorer Robert F. Scott (1868–1912), who was a close second.) McMurdo Station, described in the sidebar on page 44, is the largest base in the Antarctic.

Antarctic researchers have a lot of work to do. Paleoclimatological data indicate that parts of the Antarctic are extremely prone to rapid change in climate, and models do not always agree on what is happening, as discussed in chapter 1.

A recent expedition of the research vessel *Polarstern* to Antarctic waters measured temperature and salinity. A press release posted at EurekAlert! on April 21, 2008, noted temperatures in the Antarctic waters are presently dropping. "The Antarctic deep sea gets colder, which might stimulate the circulation of the oceanic water masses. This is the first result of the *Polarstern* expedition of the Alfred Wegener Institute for Polar and Marine Research in the Helmholtz Association that has just ended in Punta Arenas/Chile. At the same time satellite images from the Antarctic summer have shown the largest sea-ice extent on record. In the coming years autonomous measuring buoys will be used to find out whether the cold Antarctic summer induces a new trend or was only a 'slip.'"

But air temperatures are rising. Higher than normal precipitation, coupled with cooling waters, might account for the absence of sea ice melting in the Antarctic, as opposed to the rapidly vanishing sea ice in the Arctic.

Another recent study found a correlation in the West Antarctica climate with conditions prevailing in the tropical Pacific Ocean. David Schneider of the National Center for Atmospheric Research and Eric Steig at the University of Washington reconstructed West Antarctica's past climate by analyzing isotopes from ice cores obtained at eight sites during the U.S. International Trans-Atlantic Scientific Expedition, conducted from 2000 to 2002. (See chapter 1 for more information on ice cores and isotopes.) Although Antarctica has generally experienced less warming than average over the last century, Schneider and Steig discovered that West Antarctica has experienced a rise of about 1.6°F (0.9°C) in the last century, slightly higher than the global average of 1.33°F (0.74°C).

Schneider and Steig also found a relation between temperatures in West Antarctica and Pacific Ocean events. For instance, waters in the eastern Pacific Ocean warm up at irregular intervals. Such an event is called El Niño, which is Spanish for boy, a reference to Christ's birth because the phenomenon was first noticed as a warming of the waters off the coast of Peru around Christmas. (See the sidebar on page 132.) According to a National Center for Atmospheric Research news release, posted on August 12, 2008, "The data show that the Antarctic climate is highly responsive to changes in the Pacific. For example, during a major El Niño event from 1939 to 1942, temperatures in West Antarctica rose by about 6 to 10 degrees F (3-6 degrees C), and then dropped by an estimated 9 to 13 degrees F (5-7 degrees C) over the next two years." Schneider noted that this means, "As the tropics warm, so too will West Antarctica."

A FRAGILE ENVIRONMENT

The rapid warming of the Arctic and West Antarctica is already having significant effects on the environment and wildlife at the poles. Researchers have recently focused on the impact of the alarming disappearance of sea ice in the Arctic.

Despite the severe conditions, a variety of animals survive in the Arctic. The walrus, for example, is a marine mammal often seen around the latitude of the Arctic Circle. Whiskers and tusks characterize these interesting creatures, whose scientific name is *Odobenus rosmarus,* which is Latin for tooth-walking sea horse. Although walruses do not actually use their tusks for walking, the animals use these lengthy teeth, which can grow to 3.3 feet (1 m), to break holes in sea ice or haul themselves out of the water. A variety of seals, such as the hooded seal, also thrive in the Arctic.

One of the most impressive Arctic animals is the polar bear. The world's largest land carnivore, an adult male polar bear can weigh up to 1,600 pounds (720 kg). A thick coat of white fur helps insulate the animal as well as allow it to blend into its icy environment. Powerful muscles and sharp claws help the bear hunt its prey, which usually consists of seals. Polar bears typically hunt on ice. They ambush seals that surface for air in holes in the sea ice, or they use their excellent sense of smell to sniff out seals hidden underneath the icy surface. Polar bears are also superb swimmers.

The loss of Arctic sea ice will have a profound effect on all Arctic mammals. Seals and walruses use sea ice to rest and mate, and polar bears use it for hunting. Dwindling sea ice means fewer survival opportunities for these animals.

A reduction in "rest stops" will be particularly hard on the walrus. Unlike seals, which can swim for long periods of time, walruses need to pause and take breaks. The bulky animals weigh up to 3,000 pounds (1,350 kg) and require a sturdy platform, which means either terra firma or a thick slab of sea ice.

Science*Daily* posted a news release on December 10, 2007, reporting that about 40,000 walruses had recently been spotted on the Arctic coast of Russia. "According to WWF [World Wildlife Fund], this is the largest walrus haul out—areas where walruses rest when they are out of the water—registered in the Russian Arctic."

Although the large population might appear to be welcome news, scientists are worried that it is the result of migration rather than good times. As Viktor Nikiforov, director of WWF-Russia's Regional Programs, noted in the news release, "Because of climate change, ice is disappearing from the Chukchi and East Siberian seas during the summer months. This means that in the coming years new haul outs will appear along the Chukotka Arctic coast." Russian officials are attempting to protect the walruses from poachers tempted by the increase in population.

Arctic animals have adapted to their harsh environment, but evolution and adaptation are slow processes. The rapid change in climate will tax the ability of these animals to survive. If their habitats become too restricted, the density of animals will rise in the remaining livable space. As a result of this crowding, competition and fighting may further decimate the population, reducing their overall numbers. Dwindling gene pools allow for less variation and adaptation at a time when these species need more, which could doom these creatures.

MODELING THE POLAR CLIMATE CHANGE—RISING SEA LEVELS

Another important consequence of the melting of polar ice is an increase in the level of the oceans. While the melting of floating ice does not cause the sea level to rise, the melting of land ice sheets can be expected to do so. Water from melting snow and ice has to go somewhere, and the atmosphere can only hold so much water vapor at a given temperature before the vapor condenses into water. Most of the additional water will drain or fall into the ocean.

Considering the thinning and receding glaciers in Greenland and elsewhere, sea levels should have already risen. But sea level measurements are tricky. The sea does not always present a smooth surface, as any surfer knows, since waves and tides continually roil the surface. Even the contour of the ocean floor affects the water level, which is slightly higher over a seamount (undersea mountain). And although the world's oceans are connected, regional differences in sea level exist; for example, the sea level of the Pacific Ocean on the west coast of the United States is higher than that of the Atlantic Ocean on the east coast of the country. At best, researchers can gauge an average sea level based on satellite data and the measurements of tide stations along the coastline.

According to the IPCC, the average global sea level has increased by about 0.07 inches (1.8 mm) per year over the last 100 years. Some of this rise is due to melting land ice, but some of it is the result of thermal expansion. Almost all materials expand when warmed. This thermal expansion occurs because higher temperatures agitate the constituent molecules into greater motion, which increases the volume of the material. Water follows this general rule for most temperatures, except for an unusual reversal around the freezing point, where it expands with decreasing temperature due to certain molecular interactions (which is why ice has a lower density than water, and therefore floats). Global warming results in thermal expansion of the ocean as well as melting ice sheets.

One of the most critical questions at the frontier of weather and climate science involves the extent to which sea level will rise in the future. If all of the land ice on Greenland melted, for example, the added volume would swell the oceans and raise the average sea level by about 23 feet (7 m), a catastrophic increase that would flood low-lying coasts and islands. Such a drastic change would greatly alter the geography of the continents and create economic chaos.

Researchers are studying this problem by developing models of the polar climate and the effects of melting on land ice and sea level. As with all models, scientists must rely on accurate data and an adequate understanding of atmospheric and oceanic interactions. Much of the research discussed here aims to provide this information so that models will give meaningful instead of misleading results.

The IPCC extrapolated present trends and predicted that the sea level would rise about 0.6–1.94 feet (0.18–0.6 m) by the decade 2090–99. But many scientists believe this estimate is too conservative and fails to depict recent accelerations in warming and climate change. Researchers have built a number of models in an attempt to include new observations and as many of the relevant atmospheric and oceanic phenomena as possible. Some of these models suggest substantially greater increases in future sea levels. But the different models do not agree, which is often the case at the frontiers of science.

In 2008, the University of Colorado researcher W. Tad Pfeffer and his colleagues examined various scenarios of ice sheet stability and melt discharge. Publishing their report in *Science,* the researchers' model suggested limits on how fast the water could channel from melting glaciers into the ocean, constraining the rise in sea levels. "We consider glaciological conditions required for large sea-level rise to occur by 2100 and conclude that increases in excess of 2 meters [6.6 feet] are physically untenable. We find that a total sea-level rise of about 2 meters by 2100 could occur under physically possible glaciological conditions but only if all variables are quickly accelerated to extremely high limits. More plausible but still accelerated conditions lead to total sea-level rise by 2100 of about 0.8 meter [2.6 feet]."

Other researchers are more pessimistic. In a news release issued on January 16, 2009, the USGS announced the findings of a report integrating paleoclimatology data and recent changes in climate. Thirty-seven scientists from the United States, Germany, Canada, the United Kingdom, and Denmark participated in the study. One of their conclusions concerns increases in sea level. "Sustained warming of at least a few degrees (more than approximately 4° to 13°F [2.2° to 7.2°C] above average 20th century values) is likely to be sufficient to cause the nearly complete, eventual disappearance of the Greenland ice sheet. . . ." These and other changes would have a serious impact on all aspects of human society as well as Earth's wildlife and environment.

CONCLUSION

Debate in science is generally productive because it broadens perspectives and exposes ideas and theories to greater critical examination. The debate sparked by competing models will continue, and the disagreements will lead to criticism and refinement. These refinements may result in improved models that zero in on the future rise in sea level and provide a narrow range of anticipated increases. But researchers may need help to reach the necessary degree of accuracy.

An accurate prediction of sea-level change is essential if officials are to take appropriate action. Key aspects of the world's industry and economy may require regulation in order to prevent calamitous flooding. These regulations may affect many processes and systems such as those that release greenhouse gases, which would have an impact on much of the world's economy and consumption—a great deal of energy use and production at present generates these gases, as well as other potential climate-changing substances. Higher costs for electricity or transportation would put a damper on future economic growth. Underestimates of the effects of climate change will lead to inadequate preparation and possible catastrophe, but overestimates will result in wasted time and money, along with a needless crippling of the world's economy and standard of living.

Building models with higher accuracy and confidence requires more data. To obtain this data, some researchers are relying on satellites and their sky-high vantage points.

For example, the ESA attempted to launch the satellite *CryoSat* on October 8, 2005. The primary goal of *CryoSat* was to study the cryosphere, especially ice thickness. Improvements in the altimeter and antennas would have allowed researchers to make extremely precise measurements, but a rocket failure destroyed this satellite during the launch. Because of the importance of the mission, ESA immediately began work on a replacement. *CryoSat-2*, to be launched in late 2009 or early 2010, will perform the same duties for which the earlier version was intended. In February 2007, the design was reviewed and approved.

CryoSat-2 will orbit Earth at an altitude of 444 miles (717 km), making a complete revolution every 100 minutes. The orbit takes the craft over both poles, reaching latitudes of 88° (almost directly over the poles, which are at 90°). ESA anticipates the mission to last three years. On board the satellite is a sophisticated instrument called SIRAL, which stands for SAR interferometric radar altimeter (SAR refers to synthetic aperture radar). Interferometry is a technique that improves the accuracy of distance measurements by using interference of electromagnetic waves; electromagnetic waves that strike the same spot superimpose or interfere in such a manner that allows researchers to measure distances with great accuracy. The improved set of instruments makes *CryoSat-2* extremely useful for measuring ice on land and in the sea.

But sophisticated instruments must be tested and calibrated to ensure accuracy. Researchers from Denmark, Germany, the United Kingdom, and Canada have conducted expeditions to the Arctic as part of the CryoSat Validation Experiment (CryoVEx). These expeditions include flying an airplane version of the satellite's altimeter over ice fields and judging its success by comparing its readings with those taken by direct measurement. By simulating the satellite's performance, researchers can test its accuracy.

No one understands the polar climate well enough yet to predict the future, but the rapid changes are alarming. As Mark Serreze wrote in his article in *Witness the Arctic*, "We have long known that the Arctic would be the first place to see the fingerprints of greenhouse warming. This was projected in even our earliest climate models. What has caught us by surprise is the pace of change. In many ways, the projected future of the Arctic is today." Finding out what additional changes are on the horizon is a critical issue at the frontier of weather and climate.

CHRONOLOGY

1576   The English explorer Sir Martin Frobisher (ca. 1535–94) begins one of the first of many voyages by many explorers to find a hypothetical northwest passage to China around the northern portion of North America. These explorers charted parts of the Arctic, but ice blocked all attempts to locate such a passage.

1773   The British explorer Captain James Cook (1728–79) commands the first European ship known to cross the Antarctic Circle.

1830s–40s   British, French, and American naval expeditions map Antarctica's coastline.

1882–83   In the first International Polar Year, researchers engage in several expeditions into the Arctic and Antarctic.

1888	The Norwegian explorer and scientist Fridtjof Nansen (1861–1930) leads the first expedition to cross Greenland.
1903	The Scottish scientist William S. Bruce (1867–1921) leads an expedition that results in the establishment of the first permanent base in the Antarctic at Laurie Island. Argentina has maintained a presence at this base, called Orcadas, since 1904.
1909	The American explorer Robert Peary (1856–1920), along with Matthew Henson and several Inuit, report reaching the North Pole, although this claim has never been confirmed.
1911	The Norwegian explorer Roald Amundsen (1872–1928) leads an expedition that reaches the South Pole.
1928	The Australian explorer Sir George Hubert Wilkins (1888–1958) and the American aviator Carl Benjamin Eielson (1897–1929) make the first trans-Arctic Ocean flight from Point Barrow, Alaska, to Spitsbergen, Norway.
1935	The American explorer Lincoln Ellsworth (1880–1951) makes the first trans-Antarctica flight.
1955	McMurdo Station is established in the Antarctic.
1958	USS *Nautilus,* the world's first nuclear-powered submarine, makes the first undersea crossing of the North Pole.
1961	The Antarctic Treaty takes effect.
1969	The British explorer Sir Walter "Wally" Herbert (1934–2007) leads the first expedition that indisputably sets foot on the North Pole.

1970s–80s   Polar climate change begins to alarm researchers.

1983   Russia's Vostok Station in the Antarctic measures the lowest recorded temperature in the world, -129°F (-89°C).

2005   European Space Agency's *CryoSat* is destroyed when the rocket carrying it into orbit fails.

2007   Arctic sea ice shows record lows.

The fourth International Polar Year kicks off an intensive series of polar expeditions and observations.

2009   Plans call for the launch of ESA's *CryoSat-2*.

FURTHER RESOURCES

Print and Internet

EurekAlert! "The Antarctic Deep Sea Gets Colder." News release (4/21/08). Available online. URL: http://www.eurekalert.org/pub_releases/2008-04/haog-tad042108.php. Accessed July 1, 2009. Early results of the *Polarstern* expedition indicate that Antarctic waters are getting colder lately.

European Space Agency. "Arctic Sea Ice Thinning at Record Rate." News release (10/28/08). Available online. URL: http://www.esa.int/esaCP/SEMTGPRTKMF_index_0.html. Accessed July 1, 2009. Researchers at the Centre for Polar Observation and Modelling use ESA's *Envisat* to estimate Arctic sea ice thickness.

Intergovernmental Panel on Climate Change. *Climate Change 2007: Synthesis Report*. Available online. URL: http://www.ipcc.ch/pdf/assessment-report/ar4/syr/ar4_syr.pdf. Accessed July 1, 2009. IPCC scientists review climate data and models.

International Arctic Research Center. "Scientists Find Increased Methane Levels in Arctic Ocean." News release (12/15/08). Available online. URL: http://www.iarc.uaf.edu/highlights/2008/ISSS-08/. Accessed

July 1, 2009. Researcher Igor Semiletov and his colleagues discover methane concentrations are rising in Arctic waters along Siberia.

Kerr, Richard A. "Winds, Not Just Global Warming, Eating Away at the Ice Sheets." *Science* 322 (10/3/08): 33. Kerr concisely reviews two recent studies indicating that changes in ocean circulation brought about by winds are responsible for some of the melting ice sheets.

Lopez, Barry. *Arctic Dreams.* New York: Vintage Books, 2001. A reprint of a 1986 work, this book is a tribute to the natural history of the Arctic, based on the author's many trips to the region.

Mirsky, Jeannette. *To the Arctic!: The Story of Northern Exploration from Earliest Times.* Chicago, Ill.: University of Chicago Press, 1998. Mirsky relates the adventures and hardships of Arctic explorers.

Mulvaney, Kieran. *At the Ends of the Earth: A History of the Polar Regions.* Washington, D.C.: Island Press, 2001. This book describes how humans have lived and explored in the polar regions and the consequences of polar climate change.

National Center for Atmospheric Research. "Antarctic Climate: Short-Term Spikes, Long-Term Warming Linked to Tropical Pacific." News release (8/12/08). Available online. URL: http://www.ucar.edu/news/releases/2008/antarcticwarming.jsp. Accessed July 1, 2009. An analysis of West Antarctica ice cores by researchers at the National Center for Atmospheric Research and the University of Washington reveals links between the climate of West Antarctica and the tropical Pacific Ocean.

Pfeffer, W. T., J. T. Harper, and S. O'Neel. "Kinematic Constraints on Glacier Contributions to 21st-Century Sea-Level Rise." *Science* 321 (9/5/08): 1,340–1,343. The researchers report that the most likely sea-level rise from glacier melting and discharge over the next century is about 2.6 feet (0.8 m).

Science*Daily.* "Collapse of the Ice Bridge Supporting Wilkins Ice Shelf Appears Imminent." News release (4/4/09). Available online. URL: http://www.sciencedaily.com/releases/2009/04/090403080827.htm. Accessed July 1, 2009. Scientists observe dangerous weakening in the ice bridge connecting the Wilkins Ice Shelf in Antarctica to Charcot and Latady Islands.

————. "Melting Ice Displaces Walruses in the Russian Arctic." News release (12/10/07). Available online. URL: http://www.sciencedaily. com/releases/2007/11/071126143646.htm. Accessed July 1, 2009. Observers noted increases in walruses along the Russian Arctic coast, possibly due to decreases in Arctic sea ice.

————. "More Extreme Weather in the Arctic Regions." News release (2/5/09). Available online. URL: http://www.sciencedaily.com/releases/ 2009/02/090205083526.htm. Accessed July 1, 2009. Erik Kolstad at the Bjerknes Centre for Climate Research and his colleagues collected data from Arctic weather systems in order to study how vanishing sea ice may affect Arctic weather.

Serreze, Mark C. "Arctic Climate Change: Where Reality Exceeds Expectations." *Witness the Arctic* 13 (Winter 2008/2009): 1–4. Available online. URL: http://www.arcus.org/Witness_the_Arctic/winter_08_ 09/downloads/WTA_V13_No1.pdf. Accessed July 1, 2009. Serreze, a researcher at the National Snow and Ice Data Center, summarizes recent climate changes in the Arctic.

————. "Why Is the Arctic So Sensitive to Climate Change and Why Do We Care?" NOAA Arctic theme page. Available online. URL: http://www.arctic.noaa.gov/essay_serreze.html. Accessed July 1, 2009. Serreze describes the albedo feedback loops that result in the Arctic's sensitivity to warming.

United States Geological Survey. "Arctic Heats Up More Than Other Places." News release (1/16/09). Available online. URL: http://www. usgs.gov/newsroom/article.asp?ID=2109. Accessed July 1, 2009. A USGS report indicates that the Arctic is experiencing a greater than average climate change.

University of Alaska Fairbanks. "Arctic Soil Reveals Climate Change Clues." News release (10/7/08). Available online. URL: http://www. uaf.edu/news/a_news/20081007132317.html. Accessed July 1, 2009. The University of Alaska Fairbanks researcher Chien-Lu Ping and his colleagues find a previously unknown layer of organic material buried in the frozen soil in Alaska.

Wadhams, Peter. "How Does Arctic Sea Ice Form and Decay?" NOAA Arctic theme page. Available online. URL: http://www.arctic.noaa.gov/ essay_wadhams.html. Accessed July 1, 2009. Wadhams, a researcher at the University of Cambridge in the United Kingdom, reviews the formation and melting of Arctic sea ice.

Web Sites

Arctic Research Consortium of the United States (ARCUS). Available online. URL: http://www.arcus.org/. Accessed July 1, 2009. The Web site of ARCUS, a nonprofit organization devoted to Arctic research, provides information and reports on the latest discoveries.

British Antarctic Survey. Available online. URL: http://www.antarctica. ac.uk/. Accessed July 1, 2009. British Antarctic Survey conducts the United Kingdom's scientific efforts in Antarctica. The organization's Web site provides news and information on its past and present programs.

Discovering Antarctica. Available online. URL: http://www.discovering antarctica.org.uk/. Accessed July 1, 2009. This Web site, developed by the Royal Geographical Society and the British Antarctic Survey, contains descriptions, images, and videos about this frozen continent.

European Space Agency: *CryoSat-2.* Available online. URL: http://www. esa.int/esaLP/ESAOMH1VMOC_LPcryosat_0.html. Accessed July 1, 2009. ESA's *CryoSat-2,* a satellite to be launched in late 2009 or early 2010, will monitor polar ice sheets. This Web site describes the technology and science of the mission, and posts updates on the mission's progress.

National Science Foundation: Arctic Climate Research. Available online. URL: http://www.nsf.gov/news/special_reports/arctic/index. jsp. Accessed July 1, 2009. NSF funds a great deal of research in the Arctic. This Web site offers a look at some of the recent projects.

National Snow and Ice Data Center. Available online. URL: http:// nsidc.org/. Accessed July 1, 2009. The National Snow and Ice Data Center's Web site provides news and information on their data collections and latest research efforts.

3

SOLAR VARIATION AND EARTH'S CLIMATE

Bright sunshine can turn an otherwise cool day into a warm one, and on a hot day the Sun can send people scurrying into shade or air-conditioned buildings. Most of Earth's energy input comes from the Sun. Although a small amount of radioactive atoms in the planet contribute some heat, the main energy source is the Sun's electromagnetic radiation. In the Tropics, where the Sun rises high in the sky, this energy can be uncomfortably evident. As a sweaty companion once remarked to this author while hiking in the British Virgin Islands in the Caribbean, "That Sun really takes it out of you."

The Sun's input drives most processes on Earth, including weather and climate. *Solar irradiance* describes the density of the Sun's radiation that arrives at a surface, such as the surface of Earth or its upper atmosphere. (The term *solar* derives from a Latin word, *sol,* for the Sun.) Scientists sometimes refer to the average radiation from the Sun received by Earth as the *solar constant.* But the Sun's radiation output belies this term because it is at least slightly variable instead of constant. The Sun is dynamic, as can be observed from the appearance of its surface. Although it is extremely dangerous to look directly at the Sun, astronomers who use appropriately shielded instruments have noted the Sun's surface can be mottled with dark patches, a phenomenon that for the last few centuries has occurred in a fairly regular cycle.

Despite the Sun's influence on Earth's weather and climate, researchers have not paid much attention to solar variability. As noted in a fact

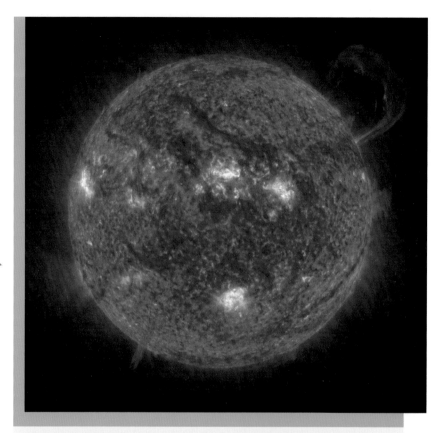

Image of the Sun taken by NASA's Extreme ultraviolet Imaging Telescope (EIT), showing hotter (white) and cooler (dark red) areas and a large prominence in the shape of a handle at the upper right *(Extreme ultraviolet Imaging Telescope Consortium/NASA)*

sheet, "The Sun and Climate," published in 2000 by the United States Geological Survey (USGS), "A direct connection between solar irradiance (solar constant) and weather and climate has been suggested for more than 100 years but generally rejected by most scientists, who assume that the effect of solar variations would be small." Researchers who study the recent warming trend and are searching for climate forcings—mechanisms that cause or force the climate to change—have focused on greenhouse gas emissions, as discussed in chapters 1 and 2.

There is no evidence for any solar variation that could have caused all or even most of the recent climate change. But the latest research suggests that the Sun's variability has been critical in past climate changes, and it may be contributing to the present situation—and possibly doing so even more strongly in the future. Disentangling the potential forcings is essential so that people can identify root causes and take appropriate action. This chapter discusses how the Sun's output varies and how this variability may affect Earth's climate.

INTRODUCTION

The Sun is an enormous ball of mostly hydrogen and helium gases. Its radius is 432,000 miles (696,000 km), with a volume that could contain 1,300,000 objects the size of Earth. Pressure and temperature in the Sun's core are so great that atomic nuclei fuse, or join, releasing large amounts of energy in the process. Much of this energy gets emitted in the form of electromagnetic radiation.

Electromagnetic radiation consists of oscillating electric and magnetic fields. The radiation's frequency can vary, from the extremely high-frequency radiation known as gamma rays to the low frequencies of radio waves. Visible light falls in the middle of the electromagnetic

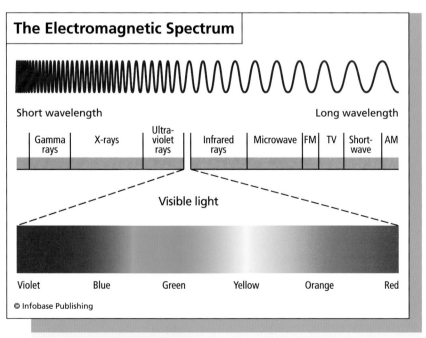

The Electromagnetic Spectrum

Short wavelength   Long wavelength

| Gamma rays | X-rays | Ultra-violet rays | Infrared rays | Microwave | FM | TV | Short-wave | AM |

Visible light

Violet   Blue   Green   Yellow   Orange   Red

© Infobase Publishing

The electromagnetic spectrum

spectrum between these two extremes, as illustrated in the following figure, and is bracketed by ultraviolet radiation, which is slightly higher in frequency, and infrared radiation, which is slightly lower. The energy of electromagnetic radiation is proportional to its frequency—the higher the frequency, the greater the energy. Gamma rays and X-rays are extremely energetic forms of radiation.

The Sun emits radiation at a variety of frequencies, but its output is highest in the visible portion of the spectrum, along with a great deal of emissions in the infrared range and a lesser amount of ultraviolet. Humans see visible light because the human eye is sensitive to electromagnetic radiation in this frequency range, an adaptation that takes advantage of the Sun's peak frequencies.

Another important consideration of radiation is the distance from the source. The Sun is roughly spherical (though it is not perfectly round), with radiation constantly leaving its entire surface. Imagine being able to follow a short interval of radiation as it left the Sun—this radiation would be a spherical shell of energy traveling at the speed of light, which is about 186,000 miles/second (300,000 km/s) in a vacuum. The surface area and volume of this shell of radiation increases as it expands into space. But the amount of energy has not changed, which means that the energy density—the energy per unit area—goes down. Any object in the path will receive a portion of this energy, but the amount depends on its distance from the source. Closer to the source, the shell has not expanded very much, so there is a lot of energy per area to strike a given object. An object of the same size but located much farther away will get less because there is less energy per unit area. The inverse square law quantifies this amount: Energy decreases with the square of the distance. An object of the same size but two times as far away from the radiation source receives one-fourth the amount of energy (the square of one-half); at three times the distance, the fraction is one-ninth.

The total solar irradiance is the amount of radiation from all frequencies that strikes the upper atmosphere of Earth. Scientists often speak of energy in terms of power, which is energy per unit time. Lightbulbs, for example, use a certain amount of electrical energy, and a typical example is a 60-watt bulb (the watt, a unit of power, honors the Scottish inventor James Watt [1736–1819]). The Sun delivers enough power to turn on a little more than two 60-watt lightbulbs for every square foot (0.09 m²) of surface in the upper atmosphere. Although this may not sound like much, the constant input, summed over the entire globe, is a huge amount. But the atmosphere reflects or absorbs some of the radiation, especially in the upper frequency range, which beneficially filters out many of the energetic and dangerous frequencies.

In its early stages, the Sun was probably much dimmer than it is now. Astronomers who study the birth and evolution of stars suspect that the Sun probably increased its brightness by about 30 percent shortly after its formation about 4.5 billion years ago. It will also change in brightness as the nuclear fuel runs out, though this event is billions of years in the future. Until then, the Sun should continue to shine at a reliable though not entirely constant intensity.

CLIMATE AND THE SUN'S RADIATION

Since solar radiation is Earth's main energy source, any significant changes should have dramatic effects on the planet's weather and climate. But these changes need not have anything to do with solar

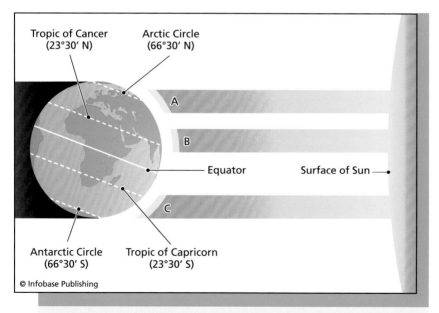

A, B, and *C* receive an identical amount of radiation, but in *A* and *C,* the radiation spreads out over a larger area. Sunlight falling on *B* is more direct and concentrated.

variability. Some of the most prominent effects are due to Earth's orientation and orbit.

The seasons, for example, regularly make their annual appearance, greatly influencing the environment as well as human activities. In the Northern Hemisphere, the months of June, July, and August are typically hot, while December, January, and February are cold. But the opposite is true in the Southern Hemisphere. The cause of these seasonal changes is a shift in Earth's orientation with respect to the Sun. Earth's axis is tilted about 23.5 degrees with respect to a line perpendicular to its orbital plane, and the tilt remains the same throughout Earth's revolution around the Sun (though the angle can vary over long periods of time). But as Earth moves in its orbit, the tilt causes one hemisphere to be pointed toward the Sun and the other hemisphere to be pointed away. The hemisphere pointed toward the Sun receives more sunlight, and the radiation strikes more directly, closer to a vertical angle. As illustrated in the figure on page 61, direct sunlight delivers more energy per unit area. The Sun rises high in the sky, and temperatures rise—summertime for this hemisphere and cool weather for the other hemisphere. The hemisphere receiving more sunlight switches when Earth reaches the other side of its orbit six months later.

Other important factors include agents that block or impede sunlight. The atmosphere reflects and filters solar radiation, and light striking at a grazing angle has to travel through more air to reach the planet's surface. Clouds reflect a great deal of light, mitigating the heat of a summer day by preventing some of the radiation from reaching the ground. An increase in certain gases or tiny dust particles suspended in the air can also exert noticeable effects. Volcanic eruptions periodically send a huge amount of gas and dust into the atmosphere and have occasionally been responsible for significant changes in weather and climate. For example, the Indonesian volcano Mount Tambora erupted in 1815, sending vast quantities of ash and gas into the air. The following year became known as the year without a summer because of the unusually cold conditions throughout much of the world—and snow in New England in June!

In addition to seasons and unanticipated changes due to geological activity, Earth has also experienced long periods of unusually cold or warm temperatures. Eras of bitter cold and encroaching ice and glaciers are known as ice ages. The last ice age ended about 10,000 to 15,000 years ago, and the present climate is a warm interglacial period. Ice ages last thousands of years, and the peaks of the last few ice ages have come at roughly 100,000-year intervals.

The regular pattern of ice ages suggests some kind of oscillatory cause. A cyclical variation in solar luminosity would account for it, but in the 1920s and '30s, the Serbian scientist Milutin Milankovitch (Milanković) (1879–1958) found an alternative explanation that is much more likely. Milankovitch computed long-term cycles in Earth's orbit. These periodic phenomena are often referred to as Milankovitch cycles. One cycle involves *eccentricity* of Earth's orbit; eccentricity is a measure of shape, with lower values being circular and higher values representing flattened, elliptical shapes. Earth's orbit slowly changes eccentricity over a 100,000-year period, going from orbits of low to high eccentricity and then back again. This cycle matches those of the ice ages. The orbit variations result in slightly different amounts of average solar irradiance due to the varying distances.

Although there is a strong correlation between eccentricity and ice age cycles, the irradiance dips are probably too small to account for all of the climate changes. But the change in irradiance in conjunction with some sort of feedback loop involving albedo or atmospheric composition are the prime suspects.

Solar variability does not cause the seasons and may not have played a role in the development of major ice ages, but researchers have discovered that the Sun undergoes periodic changes. The most prominent cycle that has been discovered to date involves the number of *sunspots.*

SUNSPOTS—"COOL" SPOTS ON THE SUN

A sunspot is a dark region on the Sun's disk. The Italian scientist Galileo Galilei (1564–1642) saw sunspots in 1610 when he somewhat recklessly trained his newly fashioned telescope on the setting Sun, and researchers have been interested in these phenomena ever since.

Sunspots appear dark because they are cooler than the surrounding *photosphere* (the outer layer of the Sun). The temperature of a large sunspot, which might be the size of a planet, is typically about 6,900°F (3,800°C), which is lower than the average temperature of 9,900°F (5,500°C). Although the spots look dark, they emit light—just not as much as the hotter surrounding regions. Sunspots form in a few days

A close-up view of the Sun's surface shows granules—cells of hot gas—and a sunspot (dark region). *(Vacuum Tower Telescope, NSO, NOAO)*

or weeks, last for several weeks or perhaps months, then disappear. The cooler temperatures appear to be the result of strong magnetic fields that reduce the flow of heat-carrying convection currents. Sunspots are also associated with active regions of the Sun, including tumultuous extensions and emissions of radiation and high-speed particles. These events create "space weather" that can temporarily elevate high-frequency radiation in space to exceedingly dangerous levels and also create magnetic and electrical disturbances on Earth.

In 1843, the German astronomer Samuel Heinrich Schwabe (1789–1875) noticed that the number of sunspots rises and falls in a cycle. This cycle has a period of about 11 years—every 11 years or so, the number of sunspots rises to a peak and then falls to a minimum. The number of sunspots last peaked in early 2002, and astronomers expect the next peak around 2012.

The American astronomer George Ellery Hale (1868–1938) discovered that the magnetism associated with sunspots flip-flops every 11 years. Magnets have two poles—north and south—that are generally stable. Hale noticed that the poles of all sunspot groups located in certain regions reverse every 11 years, switching polarity—north become south and vice versa. They reverse again in another 11 years, forming a 22-year cycle. (Earth's magnetic field also reverses poles, though irregularly and at much longer intervals on average.) This interesting relationship demonstrates the importance of magnetism in sunspot activity, although researchers do not know what causes the periodic changes in sunspots and magnetic properties. Many people refer to the cycle of sunspots and associated magnetism as the solar cycle.

The current phase of the cycle reached an extremely low minimum. A National Aeronautics and Space Administration (NASA) press release issued on April 1, 2009, provided stark numbers. On 73 percent of the 366 days of 2008, astronomers observed a "clean" Sun—no sunspots. This percentage was larger than any year since 1913. And instead of bouncing off this minimum, the Sun exhibited no sunspots on 78 of the first 90 days of 2009—87 percent of the days in the first quarter of 2009 were blank. According to the researcher David Hathaway at NASA's Marshall Space Flight Center in Alabama, "This is the quietest sun we've seen in almost a century."

The values of the maximum and minimum number of sunspots have fluctuated from cycle to cycle. Observers have made continuous records of sunspot counts since about the middle of the 19th century, but sporadic records go back to the time of Galileo. The British astronomer Edward Maunder (1851–1928) noticed that in the period from 1645 to 1715, astronomers noted only a few dozen sunspots instead of the thousands that would have been expected during normal cycles. This period in which sunspots almost disappeared is known as the Maunder minimum. The Maunder minimum coincides with one of the coldest periods of a cooling trend called the Little Ice Age, as discussed in the following sidebar.

Little Ice Age

Major ice ages have occurred periodically in Earth's recent history, and, superimposed on the main cycles, briefer trends can be found that last a few decades or a few centuries. One of these cold spells is called the Little Ice Age, which lasted from about the 15th or 16th century (there is no precise definition of the onset) until the early 19th century. Climatologists do not classify this period as a major ice age since it did not last long enough for ice sheets to expand significantly, but the people who lived through it certainly suffered from its effects.

Fishermen plying the Atlantic Ocean during the coldest times of this period reported large chunks of ice. Northern Europe suffered famines from poor crops, with growing seasons shortened by as much as a month during the harshest decades. Freezing of the Thames River in England was common during winter, and London residents celebrated with a carnival on the ice. People reported that the harbor of New York City froze in the winter of 1779–80, allowing pedestrians to stroll from Manhattan to Staten Island; the ice may even have been thick enough to permit soldiers to roll cannons. (New York City was mostly occupied by the British at this time, which came during the American Revolutionary War.)

Scientific data also provide evidence of more severe winters. Paleoclimatologists do not agree on exact start and stop dates for the Little Ice Age, but dendroclimatology—the study of tree rings and climate—shows decreased growth, especially during the 17th century through the middle of the 19th century. (See chapter 1 for more information on paleoclimatology.) The elevation at which trees can grow also decreased because of the harsher weather, and ice and snow on mountains descended to lower altitudes, sometimes encroaching upon villages and forcing residents to move. Winters were often much cooler than normal. Many areas experienced unusual shifts in precipitation and wind patterns. There were several intervals during the Little Ice Age in which the cold was particularly severe, and one of those occurred in the middle of the Maunder minimum.

Did the lack of sunspots have anything to do with the Little Ice Age? Researchers have investigated this question by studying solar variability associated with sunspots and sunspot cycles. What they have found is a slight but intriguing change in total solar irradiance.

SOLAR IRRADIANCE VARIATION

Since sunspots are cooler, darker regions of the Sun, an increase in their number might be expected to reduce the Sun's brightness, thereby decreasing the amount of sunlight Earth receives. Although this argument is logical, it is not correct. This somewhat surprising result is one of the reasons why scientists insist on performing measurement rather than relying on logic—nature does not always do what it is expected to do.

Prior to 1978, scientists wanted to search for solar variations but could not do so accurately. Although researchers could measure the amount of sunlight reaching Earth's surface, this value fluctuates due to clouds as well as varying numbers of particles suspended in the atmosphere. A fluctuation in the energy reaching Earth's surface could be due to solar variation or it could be due to weather or atmospheric conditions on the planet.

However, on October 25, 1978, the satellite *Nimbus-7* lifted off from Vandenberg Air Force Base in California. (Nimbus is a type of cloud.) Operated by the National Oceanic and Atmospheric Administration (NOAA) and NASA, *Nimbus-7* was one of a series of weather satellites designed to perform atmospheric and radiation measurements. One of the instruments aboard the craft was called Earth Radiation Budget (ERB) and contained sensors to record the level of solar radiation striking Earth's atmosphere. This and subsequent orbiting instruments have given scientists the means to gauge solar irradiance in the absence of interference from atmospheric variables.

Satellite data measures solar power in watts, as mentioned above. This data indicates that the total solar irradiance is 127 watts per square foot (1,370 W/m^2). Scientists have now accumulated satellite measurements over several solar cycles that show a correlation between solar irradiance and the number of sunspots. Solar irradiance changes by about 0.1 percent during the course of the cycle and is greater when sunspots are more numerous. The reason for this is that although sunspots are darker, the surrounding regions grow a little brighter. In addition, increased solar activity

Solar Radiation and Climate Experiment (SORCE)

Launched on January 25, 2003, by a Pegasus XL rocket, the *Solar Radiation and Climate Experiment* mission continues the monitoring of solar radiation begun with *Nimbus-7*. The satellite orbits at an altitude of 400 miles (645 km), which is well above most of Earth's atmosphere. Every satellite soars beyond most of the atmosphere in order to preclude air resistance that would slow the craft down and yank it from its orbit, but in *SORCE's* case, the altitude is also needed in order to make irradiance measurements outside of the varying influence of the atmosphere. The Laboratory for Atmospheric and Space Physics at the University of Colorado currently manages the project.

One of the *SORCE* instruments is a total irradiance monitor, which measures total solar irradiance with a special kind of *radiometer*. This device gets exposed to sunlight for a certain portion of the time and measures the power. But the instruments and electronics of the satellite itself also generate heat, some of which could warm the radiometer and throw off the measurement. To avoid this error, the radiometer detects the internal energy input during dark periods, when it is not exposed to sunlight, and subtracts this value from the total. The total irradiance monitor can measure solar irradiance with a precision of about 0.035 percent.

Space probes and satellites have limited power sources and can function for only a number of years, and there is only so much money to maintain the staff necessary to operate it and collect the data. Project managers originally scheduled *SORCE* to terminate in 2008, but because of the importance of the mission, the end date was extended to 2012. This allows *SORCE* to gather data during at least part of the time when the Sun reaches the next maximum in the solar cycle.

results in slightly greater output in certain portions of the spectrum, such as ultraviolet.

An important satellite launched in 2003 has contributed even more to the study of solar variability and its possible impact on Earth's climate. The *Solar Radiation and Climate Experiment (SORCE)* is the subject of the sidebar on page 68.

The change of 0.1 percent over the solar cycle is not large in relative terms—only one part in 1,000. But Earth's climate is complex, with many interactions that can amplify small changes. One example was discussed in chapter 2, where slightly higher temperatures in the Arctic begin to melt snow and ice, decreasing surface reflectivity and allowing more energy to be absorbed, which causes more warming, and so on. And perhaps the Sun would have experienced even greater variability during the Maunder minimum, which may have resulted in far less radiation than normal since solar irradiance decreases during sunspot minimums.

Charles A. Perry of USGS and Kenneth J. Hsu of Tarim Associates in Zurich, Switzerland, constructed a climate model that suggested solar variations can have important effects. In their report, "Geophysical, Archeological, and Historical Evidence Support a Solar-Output Model for Climate Change," published in a 2000 issue of the *Proceedings of the National Academy of Sciences,* the researchers used a variety of data sources to compare with the results of the model, including the minor ice ages. Climate forcing was due to solar variation occurring in a cycle composed of some multiple of 11 years. The researchers wrote, "Although the cold periods of the little ice ages vary in length and severity, they seem to track the solar-output model reasonably well."

RECENT CLIMATE CHANGE AND THE SUN

In view of the possible influence of solar variability on Earth's climate, some people have wondered if the Sun is playing a role in recent climate change. As discussed in the previous two chapters, researchers have documented a general warming trend in the last century, along with retreating ice sheets and vanishing sea ice in the Arctic. According to the Intergovernmental Panel on Climate Change (IPCC), the average surface temperature has risen about 1.33°F (0.74°C) in the last 100 years.

IPCC scientists believe increased emissions of greenhouse gases (GHGs) are to blame for the recent warming. In IPCC's most recent assessment, "Climate Change 2007," scientists noted that "Changes in the atmospheric concentrations of GHGs and aerosols, land cover and solar radiation alter the energy balance of the climate system and are drivers of climate change. They affect the absorption, scattering and emission of radiation within the atmosphere and at the Earth's surface. The resulting positive or negative changes in energy balance due to these factors are expressed as radiative forcing. . . . Human activities result in emissions of four long-lived GHGs: CO_2 [carbon dioxide], methane (CH_4), nitrous oxide (N_2O) and halocarbons (a group of gases containing fluorine, chlorine or bromine)."

The onset of the Industrial Revolution greatly elevated the emissions of greenhouse gases. According to the IPCC, "Global atmospheric concentrations of CO_2, CH_4 and N_2O have increased markedly as a result of human activities since 1750 and now far exceed pre-industrial values determined from ice cores spanning many thousands of years." The panel has "very high confidence that the global average net effect of human activities since 1750 has been one of warming, with a radiative forcing of +1.6 [+0.6 to +2.4] W/m^2." (The value 1.6 is the best estimate, with a range of 0.6–2.4; 1.6 W/m^2 = 0.15 W/ft^2.) "In comparison, changes in solar irradiance since 1750 are estimated to have caused a small radiative forcing of +0.12 [+0.06 to +0.30] W/m^2, which is less than half the estimate given in the TAR [the IPCC's previous assessment]" (0.12 W/m^2 = 0.011 W/ft^2).

Many researchers agree that GHGs have been far more important than solar variability in recent climate change. In a review article on the subject in a 2006 issue of *Nature*, Peter Foukal of Heliophysics, Inc., and his colleagues wrote, "The variations measured from spacecraft since 1978 are too small to have contributed appreciably to accelerated global warming over the past 30 years."

But how much the solar cycle and its effect on irradiance contribute to climate change is not yet completely determined. Charles D. Camp and Ka Kit Tung of the University of Washington compared surface temperatures during the period 1959 to 2004 with the phase of the solar cycle. In 2007, they published their findings in *Geophysical Research Letters*. They reported that during maximum solar activity the average global temperature rose 0.36°F (0.2°C).

This increase due to the solar cycle, should it be supported by additional research, is only a fraction of the 1.33°F (0.74°C) increase reported by IPCC for the last 100 years, but it is not negligible and is more significant than IPCC scientists reported in *Climate Change 2007*. Other researchers are also finding a larger portion of recent climate change may be due to the Sun. For example, Manuel Vázquez, a researcher at the Canary Islands' Astrophysics Institute, spoke at a 2008 conference on the Sun and climate change sponsored by Complutense University in Madrid, Spain. On July 18, 2008, Science*Daily* posted a news release summarizing Vázquez's opinion that solar activity may account for about 15 to 20 percent of present climate changes. But Vázquez did not challenge the primary importance of human activity.

The debate continues, partly caused by the difficulty in interpreting measurements. As Foukal and his colleagues noted in their *Nature* paper, "Additional evidence on past TSI [total solar irradiance] variations might, in principle, be gleaned from reconstructions of the Earth's climate record. But the relationship between TSI variations and climate is complex, being significantly affected by uncertainties in the climate sensitivity, the large thermal inertia of the oceans, and the effects of non-solar forcings such as volcanism and human-induced factors." More research is needed.

Even though the solar cycle's contribution to recent climate change is not yet fully understood, it does not seem likely to account for a large percentage of it. However, a few researchers are beginning to wonder if longer-term variations in the Sun's output may be playing a more important role than previously suspected.

LONG-TERM VARIATIONS IN SOLAR LUMINOSITY

The solar cycle—the periodic waxing and waning of sunspots and the associated magnetic variations—is firmly established. This cycle is the only one in which researchers have convincing evidence, and it may be the only one that exists. Foukal and his colleagues gave their skeptical assessment of the possibility of long-term variations in their *Nature* paper: "Overall, we can find no evidence for solar luminosity variations of sufficient amplitude to drive significant climate variations on centennial, millennial and even million-year timescales."

But scientists have only been able to measure solar irradiance since 1978. This is enough time to cover several of the sunspot cycles, which only last slightly more than a decade, but it may not be enough time to detect variations with longer periods. And as Foukal and other scientists have noted, studying solar variations from the traces that may have been left in Earth's past is subject to a number of different interpretations.

Despite the difficulty of this kind of research, the Maunder minimum indicates that some sort of solar variability may take place on a much longer timescale than the well-known solar cycle. The model of Perry and Hsu, described on page 69, suggests that this waxing and waning of the solar cycle's magnitude can strongly affect Earth's climate. If the Sun exhibits this kind of activity, solar variability may be an important component in Earth's climate.

This issue is a contentious one. If human activity is mostly to blame for climate change, as IPCC and many other scientists claim, then constraining industrial and technological emissions of GHGs would make sense. But much of society relies on these processes, which produce the majority of energy and transportation services, especially in economically developed countries such as the United States. Changes will probably be expensive and could have adverse economic consequences, at least over the short term, and some industries would be more seriously affected than others. The causes of global warming have become a political issue as much as a scientific one in recent U.S. elections.

Some scientists believe that media coverage—newspapers, magazines, and television—have unfairly portrayed the debate as over. In a 2008 issue of *Physics Today,* the Duke University researcher Nicola Scafetta and U.S. Army Research Office scientist Bruce West wrote, "The causes of global warming—the increase of approximately 0.8±0.1°C in the average global temperature near Earth's surface since 1900—are not as apparent as some recent scientific publications and the popular media indicate. We contend that the changes in Earth's average surface temperature are directly linked to two distinctly different aspects of the Sun's dynamics: the short-term statistical fluctuations in the Sun's irradiance and the longer-term solar cycles. This argument for directly linking the Sun's dynamics to the response of Earth's climate is based on our research and augments the interpretation of the causes of global warming presented in the United Nations 2007 Intergovernmental Panel on Climate Change (IPCC) report."

But while no one disputes the existence of the sunspot cycle, the long-term variations are controversial. Attempts to resolve this issue will require paleoclimatology studies.

One recent effort involved tree rings. Trees grow each year, becoming thicker as the annual rings add to their girth (see the figure on page 14). As they grow, trees incorporate molecules made of carbon and other elements. Some of the carbon is in the form of a particular isotope, carbon-14, which is radioactive and eventually decays. Atoms of carbon-14 arise when high-speed cosmic rays, which are mostly particles such as protons, induce nuclear transformations in the upper atmosphere; some of the carbon-14 atoms drift to the surface, where they participate in the same reactions as other carbon isotopes. (See the sidebar on page 15.)

Scientists use the decay of carbon-14 to age organic (carbon-containing) materials, but the concentration of carbon-14 in the air depends on the number of cosmic rays. This number depends on the state of the Sun, although the relationship is an inverse one—when the Sun is more active, cosmic rays go down. The reason is that the Sun's magnetic field increases during active periods, which deflects cosmic rays—which are mostly charged particles—preventing them from reaching Earth's atmosphere. When the Sun is more active, fewer carbon-14 atoms are produced.

Sami K. Solanki, a scientist at the Max Planck Institute for Solar System Research, and his colleagues studied the carbon-14 in tree rings to reconstruct the past 11,400 years of solar activity. The researchers located the buried remains of old trees and analyzed their carbon-14 content. This analysis provided a year-by-year record, which was then dated and placed in the appropriate chronological order. In this way the scientists acquired a history of past solar activity. Solanki and his colleagues published their findings in a 2004 issue of *Nature.*

The findings were surprising—the Sun has been more active recently than at any time in thousands of years. "According to our reconstruction, the level of solar activity during the past 70 years is exceptional, and the previous period of equally high activity occurred more than 8,000 years ago." This high level of activity corresponds with a greater than usual number of sunspots—the opposite of the Maunder minimum, which occurred during the Little Ice Age. According to this study, the Sun seems to have been bathing Earth with more radiation lately.

While these results hint at a possible solar contribution to global warming, the activity increase is relatively small. Solanki and his colleagues wrote, "Although the rarity of the current episode of high average sunspot numbers may indicate that the Sun has contributed to the unusual climate change during the twentieth century, we point out that solar variability is unlikely to have been the dominant cause of the strong warming during the past three decades."

SUN AND CLIMATE CONNECTIONS

Tree rings, ice cores, and the concentrations of certain isotopes provide hints, but the measurements are difficult and the data reveals more about the climate than the Sun—separating the solar component is not always possible. In the search for more data on the long-term connections of the Sun and climate, some researchers have turned to ancient history.

Ancient civilizations such as the Sumerians in Mesopotamia and the Egyptians thrived when harvests were plentiful—and starved when they were not. A few years of inadequate water supply or cooler temperatures could mean the difference between life and death. Because of their reliance on the weather and climate, astute observers in these civilizations studied weather patterns and sought clues that would allow them to predict the future. Part of this effort included keeping accurate records of past conditions. Archaeologists have discovered some of these records, providing glimpses into ancient cultures as well as the weather conditions in which they lived.

Alexander Ruzmaikin and Joan Feynman, researchers at NASA's Jet Propulsion Laboratory in California, and Yuk Yung at the California Institute of Technology recently studied surviving Egyptian records of the water levels of the Nile River. Annual flooding of the Nile River was critical to Egypt's agriculture, so Egyptians kept a close watch on the river's level. Most of the surviving documents are sketchy and incomplete, but there is a period between 622 and 1470 C.E. in which records of the water levels at a certain point in the Nile are fairly thorough. The researchers concentrated on this 848-year period. The data provide a lot of information on Earth's climate at the time because of the Nile's im-

Aurora borealis in Alaska *(U.S. Air Force, photo by Senior Airman Joshua Strang)*

portance in Africa—the river drains about 10 percent of the continent, and its two main sources (Lake Tana in Ethiopia and Lake Victoria in parts of Tanzania, Uganda, and Kenya) are near the equator and reflect conditions in the Atlantic and Indian Oceans.

The Nile water levels provided data on climate, but what the researchers needed was a way to compare this data to the Sun's activity. To obtain information on solar activity during this time, the researchers examined records of auroras kept by peoples in northern Europe and the Far East. Auroras, or northern lights, are bright lights appearing in the night sky around or near the Arctic (a similar phenomenon occurs in the Southern Hemisphere). Long ago, people often interpreted lights in the sky as portents or omens, and some cultures kept accurate records of their occurrence. The Sun's emission of charged particles creates auroras because Earth's magnetic field deflects the particles, which bump and collide their way toward the polar regions and emit light. Aurora records give researchers a measure of solar activity.

A news release issued by the Jet Propulsion Laboratory on March 19, 2007, announced the findings. Ruzmaikin and his colleagues discovered a link between solar activity and Earth's climate, as revealed by these written records. "The Nile water levels and aurora records had two somewhat regularly occurring variations in common—one with a period of about 88 years and the second with a period of about 200 years." This relationship suggests that solar activity influences important aspects of Earth's climate.

The researchers do not yet know how the variations in the Sun's activity may have caused the climate fluctuations. Ruzmaikin and his colleagues think one possibility is ultraviolet radiation, the variation of which may have an impact on certain weather patterns. One of these patterns involves air circulation over the Atlantic and Indian Oceans, which affects temperature and precipitation across broad regions, including equatorial Africa. Increased solar activity may result in a significant reduction of precipitation through a complex series of changes.

Another group of researchers also studied ancient history, but in this case the records were not written but came from layered deposits in a stalagmite found in the Wanxiang Cave in northern China. A stalagmite is a formation that rises from the floor of a cave as it accumulates minerals from water dripping from the ceiling. The paleoclimatologist Pingzhong Zhang at Lanzhou University in China and a large team of researchers analyzed isotopes taken from samples of the stalagmite. The researchers dated the deposits by measuring the ratios of uranium and thorium, the radioactive decay of which provides a steady "clock." Variations in oxygen isotopes in the sample gave an indication of the amount of rain in the region (see chapter 1), and the uranium-thorium "clock" let the scientists figure out when these rainfall variations took place. With these data, Zhang and the research team pieced together a history of rainfall in northern China for the past 1,810 years. This rain reflects the activity of the Asian monsoon, an important climate pattern that brings wet weather in the summer.

The researchers discovered certain variations in the amount of rain suggested a solar influence, and they published their findings in a 2008 issue of *Science*. For example, there was a roughly 10.5-year pattern that closely matched the solar cycle. Rain variability also sometimes corresponded with records of solar activity, although the correspondence was not strong. And the researchers noted, "The sign of the correlation between the AM [Asian Monsoon] and temperature switches around 1960, suggesting that anthropogenic [human-made] forcing superseded natural forcing as the major driver of AM changes in the late 20th century." This means that, according to this study, factors arising from human activities, such as GHG emissions, have been more important lately in determining the amount of rain than other factors, such as solar variability.

CONCLUSION

The study of solar variability and its impact on Earth has yielded valuable clues about the nature of weather and climate. Solar radiation is the primary energy source for the planet, and fluctuations in its amount or composition can have significant effects on climate. Cycles and variations that researchers have discovered thus far in solar output have been small, but increasingly strong evidence supports the idea that Earth's climate has adjusted accordingly.

Predictions of future climate have become extremely important recently due to the observed climate changes. Making an accurate prediction requires knowledge of all the important factors and their influence. One of these factors—the Sun and its variability—has only been recently discovered, and the extent of its influence has yet to be determined. Climate specialists still have much work to do.

More data along with improved instrumentation and wider coverage will help researchers meet their goals. As is the case with much research these days, satellites will play a major role. To fulfill the growing demands, space agencies such as NASA and the European Space Agency (ESA) are developing advanced satellites and onboard instruments. In 2009, NASA plans to launch *Glory* to study climate change. One of the satellite's two main objectives is to measure the properties and distributions of *aerosols* and their effects on climate. The other primary objective is the continuation of the close monitoring of the total solar irradiance that began with *Nimbus-7* in 1978 and is currently performed by *SORCE* (which is scheduled for shutdown in 2012). *Glory* will house a new version of *SORCE*'s total irradiance monitor in order to carry on this effort. The altitude of the satellite will be 437 miles (705 m), slightly higher than *SORCE*.

Another launch scheduled for 2009 is NASA's *Solar Dynamics Observatory*. Researchers have designed this satellite to probe the causes of the Sun's variability and how this variability affects Earth. Three different instrument assemblies will monitor specific aspects of the Sun simultaneously. The Helioseismic and Magnetic Imager (HMI) will measure motion in the Sun's photosphere and how processes inside the Sun are related to the magnetic field. The Atmospheric Imaging Assembly (AIA) will study fluctuations in the outer layers of the Sun, including flares—intense variations in brightness. And the Extreme Ultraviolet Variability Experiment (EVE) will measure ultraviolet emissions with high precision. The satellite's instruments will take an image of the full disk of the Sun every 10 seconds. Scientists expect this ambitious project to generate a huge amount of data, so mission specialists decided to put the satellite in an orbit that facilitates communication with the ground in order to maximize data transfer rates. The satellite will be in a geosynchronous orbit—hovering about 22,200 miles (35,800 km) over a fixed point on Earth's surface—and in constant contact with a ground station located in New Mexico.

These projects, and many others, should provide some of the data needed for researchers to gain a better understanding of how solar variability affects Earth's climate. While there is no firm evidence that solar variations are responsible for most of the recent climate change, correlations of these variations with shifting weather patterns on Earth and important events such as the Little Ice Age suggest a major role in climate. These variations can strike again at any time in the future. Although another Little Ice Age seems unlikely any time soon, no one can be certain until scientists have uncovered the Sun-climate links.

Another critical aspect of this research is the light it can shed on global climate change. Global climate change affects everyone, as will any decisions on how to fight it, and the causes of global climate change must be established and quantified objectively and fairly. Facts and judgments should come from rigorous experiments and theories—most citizens are comfortable with decisions based on science. The efforts of scientists at the frontiers of weather and climate will be critical in this endeavor.

CHRONOLOGY

| 1610 | The first person to notice a sunspot will probably remain forever anonymous, but the Italian scientist Galileo Galilei (1564–1642) was one of the earliest observers to view a sunspot with a telescope. |

1801	The German-British astronomer William Herschel (1738–1822) speculates that variations in the Sun could cause climate change.
1843	The German astronomer Samuel Heinrich Schwabe (1789–1875) discovers the 11-year period of the sunspot cycle.
1890s	The British astronomer Edward Maunder (1851–1928) discovers that observers of the period from 1645 to 1715 reported exceptionally few sunspots.
1914	The American geographer Ellsworth Huntington (1876–1947) speculates that sunspots may affect Earth's climate.
1920s	The Serbian scientist Milutin Milankovitch (1879–1958) proposes that periodic variations in Earth's orbit are major contributors to the development of ice ages.
1925	The American astronomer George Ellery Hale (1868–1938) and his colleagues identify a 22-year cycle of solar magnetism.
1978	*Nimbus-7,* the first satellite to continuously monitor total solar irradiance, launches.
1995	Using data accumulated from satellite observations, astronomers discover that solar irradiance varies about 0.1 percent over its 11-year cycle.
2002	The solar cycle reaches a maximum.
2003	The *Solar Radiation and Climate Experiment (SORCE)* begins.
2007	The Intergovernmental Panel on Climate Change (IPCC) releases the report *Climate Change 2007,* its fourth major assessment of the situation.

2008–09	Astronomers report the Sun is in an extremely quiet period.
2009	Expected launch of NASA satellite *Glory*.
	Expected launch of NASA's *Solar Dynamics Observatory* mission.
2012	Expected peak in the solar cycle.

FURTHER RESOURCES

Print and Internet

Camp, C. D., and K. K. Tung. "Surface warming by the solar cycle as revealed by the composite mean difference projection." *Geophysical Research Letters* 34 (2007): L14703, doi: 10.1029/2007GL030207. Accessed July 1, 2009. The researchers compared surface temperatures during maximum and minimum periods of the solar cycle and found a small difference of 0.36°F (0.2°C).

Exploratorium: The Museum of Science, Art and Human Perception. "Sunspots." Available online. URL: http://www.exploratorium.edu/sunspots/. Accessed July 1, 2009. This excellent series of articles describes what scientists have learned about sunspots. A short discussion of the possible impact of the solar cycle on Earth's climate is included.

Fagan, Brian. *The Little Ice Age.* New York: Basic Books, 2000. This book tells the story of cold winters and severe hardships endured during the late Middle Ages and Renaissance, affecting much of Europe and North America.

Foukal, P., C. Fröhlich, H. Spruit, and T. M. L. Wigley. "Variations in Solar Luminosity and Their Effect on the Earth's Climate." *Nature* 443 (9/14/06): 161–166. The researchers review the latest findings on solar variability and the climate.

Golub, Leon, and Jay M. Pasachoff. *Nearest Star: The Surprising Science of Our Sun.* Cambridge, Mass.: Harvard University Press, 2002. The authors, both astronomers, provide an accessible introduction to solar science.

Intergovernmental Panel on Climate Change. *Climate Change 2007: Synthesis Report.* Available online. URL: http://www.ipcc.ch/pdf/assessment-report/ar4/syr/ar4_syr.pdf. Accessed July 1, 2009. IPCC scientists review climate data and models.

Jet Propulsion Laboratory. "NASA Finds Sun-Climate Connection in Old Nile Records." News release, (3/19/07). Available online. URL: http://www.jpl.nasa.gov/news/features.cfm?feature=1319. Accessed July 1, 2009. Scientists discover correlations in records of Nile River water levels and observations of auroras in northern Europe and the Far East.

National Aeronautics and Space Administration. "How Low Can It Go? Sun Plunges into the Quietest Solar Minimum in a Century." News release, (4/1/09). Available online. URL: http://www.nasa.gov/topics/solarsystem/features/solar_minimum09.html. Accessed July 1, 2009. NASA announces statistics showing that the current solar cycle is at an extremely low minimum.

Perry, Charles A., and Kenneth J. Hsu. "Geophysical, Archeological, and Historical Evidence Support a Solar-Output Model for Climate Change." *Proceedings of the National Academy of Sciences* 97 (2000): 12,433–12,438. The researchers report on a model that suggests solar variations can have important climate effects.

Scafetta, Nicola, and Bruce J. West. "Is Climate Sensitive to Solar Variability?" *Physics Today* 61 (March 2008): 50–51. The researchers argue that solar variability is an important factor in recent climate changes.

Science*Daily.* "Sun Could Cause 15% to 20% of Effects of Climate Change, Researcher Says." News release (7/18/08). Available online. URL:http://www.sciencedaily.com/releases/2008/07/080717224333.htm. Accessed July 1, 2009. This news release summarizes the findings of Manuel Vázquez, a researcher from the Canary Islands' Astrophysics Institute, on possible solar contributions to the recent warming trend.

Solanki, S. K., I. G. Usoskin, B. Kromer, M. Schüssler, and J. Beer. "Unusual Activity of the Sun during Recent Decades Compared to the Previous 11,000 years." *Nature* 431 (10/28/04): 1,084–1,087. The researchers reconstructed the history of solar activity and found that it has recently increased.

Soon, Willie Wei-Hock, and Steven H. Yaskell. *The Maunder Minimum and the Variable Sun-Earth Connection.* Singapore: World Scientific Publishing, 2003. This book offers a detailed discussion of sunspots, the solar cycle, and possible links with Earth's climate.

United States Geological Survey. "The Sun and Climate." Fact Sheet FS-095-00 (August 2000). Available online. URL: http://pubs.usgs. gov/fs/fs-0095-00/fs-0095-00.pdf. Accessed July 1, 2009. This report summarizes research on solar variability and climate change.

Zhang, Pingzhong, Hai Cheng, R. Lawrence Edwards, Fahu Chen, Yongjin Wang, Xunlin Yang, et al. "A Test of Climate, Sun, and Culture Relationships from an 1810-Year Chinese Cave Record." *Science* 322 (11/7/08): 940–942. Using layered deposits in a cave, the researchers found correlations in rainfall and solar variability.

Web Sites

Marshall Space Flight Center: Solar Physics. Available online. URL: http://solarscience.msfc.nasa.gov/. Accessed July 1, 2009. The Marshall Space Flight Center is a NASA research center in Huntsville, Alabama. This Web site contains news and information on the center's Solar Physics Group, which has been conducting research on the Sun since the early 1970s.

NASA: *Glory.* Available online. URL: http://glory.gsfc.nasa.gov/. Accessed July 1, 2009. News and information can be found here on the NASA satellite whose mission is to study aerosols and monitor solar irradiance.

NASA: *Solar Dynamics Observatory.* Available online. URL: http://sdo. gsfc.nasa.gov/. Accessed July 1, 2009. This Web site offers news and information on the satellite designed to study solar variability.

Solar Radiation and Climate Experiment (SORCE). Available online. URL: http://lasp.colorado.edu/sorce/. Accessed July 1, 2009. This Web site contains news and information on the satellite mission that addresses solar variability and climate change.

4

TORNADO FORMATION

On March 14, 1925, a tornado touched down in southeastern Missouri. Residents of nearby Ellington and other communities scrambled for cover, probably expecting that the storm would dissipate before long—most tornadoes last less than about 10 minutes. This tornado quickly passed Ellington, moving toward Illinois with a speed of about 70 MPH (112 km/hr), but it lasted a lot longer than most tornadoes. Wind speed was not recorded, but researchers later estimated the winds were blowing at greater than 250 MPH (400 km/hr), exerting a devastating force on everything in its path. The tornado became known as the Tri-State Tornado because it swept a continuous path more than 219 miles (350 km) long through three states—Missouri, Illinois, and Indiana. In its wake lay the ruins of towns such as Annapolis, Missouri, Murphysboro, Illinois, and Griffin, Indiana. Fatalities totaled 695, making the Tri-State Tornado the deadliest tornado in U.S. history.

The disaster prompted meteorologists and scientists to launch serious efforts to understand more about these ferocious storms. A tornado is a rapidly rotating column of air that extends from a cloud to the ground. Most tornadoes have wind speeds exceeding 50 MPH (80 km/hr), and the record is 318 MPH (509 km/hr), measured for an Oklahoma tornado on May 3, 1999 (this is the fastest documented wind speed on Earth). The United States is especially prone to violent storms. Tornadoes have been observed in all inhabited areas of the world, but meteorologists detect about 1,000 tornadoes every year in the United States—a rate of about 10 times higher than in Canada, which experiences the second most number. In terms of tornadoes per unit area, the United Kingdom and the Netherlands are the leaders,

Tornado in an early stage of development at Union City, Oklahoma *(NOAA Photo Library, NOAA Central Library; OAR/ERL/National Severe Storms Laboratory [NSSL])*

though the wind speeds of their storms are generally much weaker than the major storms of the United States.

Although researchers have been studying tornadoes for decades, the power and violence of these storms limit the measurements that can be made. Scientists have learned much about the conditions under which tornadoes form, but they still cannot predict when and where the next one will occur or the path it may follow. Investigators must monitor broad areas and race to a developing storm in order to study the process. This chapter describes what researchers know about the formation of tornadoes and how this knowledge has led to better warning systems that save lives. Additional research at this frontier of science will help improve weather forecasting to an even greater extent, as well as contributing to the satisfaction of understanding one of the most intense weather systems on the planet.

INTRODUCTION

The term *tornado* derives from a Spanish word, *tronada,* meaning *thunderstorm.* Tornadoes are also known as twisters or whirlwinds. They form out of rotating winds, which often begin when a storm, such as a thunderstorm, develops a *vortex*—a spinning mass of air. A vortex called a *mesocyclone* may develop, most often in a region of low pressure in the storm. If this circulatory wind pattern extends from the cloud onto the surface and creates a funnel from cloud to ground, a tornado is born. The funnel is sometimes not visible since it consists of rotating air. But usually the funnel can be seen because of the presence of various agents including dust particles and water vapor, which condenses and becomes cloudy during the violent motion. The effects of the funnel are always obvious at the ground, where the winds kick up plenty of dirt and debris.

Tornadoes come in various shapes and sizes. Some look like ropes, while others are much wider; the diameter can vary from about 5 feet (1.5 m) to 1 mile (1.6 km), but many are around 250 feet (76 m). The size does not necessarily reflect its wind speed or the potential damage a tornado may cause.

Most large tornadoes rotate in a specific direction. In the Northern Hemisphere, these tornadoes tend to rotate counterclockwise (as viewed from above), and in the Southern Hemisphere they rotate in the opposite direction. The reason for this spin preference is discussed in the section Tornadogenesis—Birth of a Twister on page 94. Smaller tornadoes, however, may not follow this pattern.

There are several types of tornadoes. When a tornado consists of two or more columns, it is called a multivortex tornado. A *waterspout* is a tornado that occurs over water. Sometimes a tornado may form in the absence of thunderstorms and mesocyclones, in which case it is known as a *landspout.*

Other wind circulations include the weaker but more familiar spinning column of air and dust that is often called a dust devil or sun devil. They tend to form during hot days, when sudden gusts of wind twist a warm column of rising air. Most of them are weak and dissipate within a few minutes without causing any harm.

Categorizing tornadoes by intensity can be difficult because on many occasions meteorologists may not have had the means to measure the wind speed of a short-lived storm. This was particularly true in 1971, when the Japanese physicist Tetsuya Theodore "Ted" Fujita

(1920–98) developed a scale to classify tornadoes based on the damage they left in their wake, as discussed in the following sidebar. The Fujita scale assigns numbers from F0, which refers to the weakest tornado, to F5, the strongest. In 2007, meteorologists began using the enhanced Fujita scale, which has the same basic scale but has been revised based on data accumulated since Fujita's original work.

Tornadoes can last as little as a few seconds or as long as a few hours. Storms often spawn a series of tornadoes, one after another, making it difficult to gauge the duration of any one twister. But the Tri-State Tornado of 1925 left a continuous track, which makes its 3.5-hour duration likely due to a single tornado.

People who have survived the passage of a severe tornado often speak of the wind as making a roaring sound, as if coming from a train.

Fujita Scale—Ranking Tornadoes

As a young physicist, Ted Fujita had a tragic opportunity to study an unprecedented amount of wind and blast damage. When he was 24 years old, Fujita went to Hiroshima and Nagasaki to analyze the effects of the atomic bombs that had been dropped on those cities during World War II. In 1953, Fujita joined the faculty of the University of Chicago in Illinois, where he applied his analytical skills to the study of tornadoes and the damage they leave behind. In 1971, he published his scale, called the Fujita scale, which is based on a tornado's damage. The ranks are F0 (light damage), F1 (moderate), F2 (considerable), F3 (severe), F4 (devastating), and F5 (total). The Tri-State Tornado of 1925 preceded the scale, but researchers have since estimated that it was an F5. Most tornadoes fall into the F0 or F1 ranking.

Although the scale is based on damage, Fujita calculated the wind speed that he believed would correspond with a tornado's effects. Subsequent data has indicated that these calculations were not very accurate. As a result, researchers

The National Weather Service (NWS), a branch of the National Oceanic and Atmospheric Administration (NOAA), monitors weather conditions and in the appropriate circumstances issues tornado watches, which means that a tornado is possible, and tornado warnings, which means that one has been observed or expected to develop. People in threatened areas should proceed to the lowest floor of the building, preferably to a windowless room, and find shelter under a heavy table or mattress. (Long ago, some homeowners thought that opening windows to equalize air pressure would help save their homes, but this has not proven useful and can be dangerous if the occupant is around a window when the wind blows it out.) Attempting to drive through a tornado or staying in a light-weight structure such as a mobile home is not a good idea because the high winds can easily flip it over. The NWS advises

in 2007 introduced the Enhanced Fujita Scale, which assigns ranks EF0–EF5. The scale continues to be based on damage, but the corresponding wind estimates have been updated:

- EF0, wind speed estimated at 65–85 MPH (104–136 km/hr)
- EF1, 86–110 MPH (137–176 km/hr)
- EF2, 111–135 MPH (177–216 km/hr)
- EF3, 136–165 MPH (217–264 km/hr)
- EF4, 166–200 MPH (265–320 km/hr)
- EF5, >200 MPH (320 km/hr)

Despite the revisions, Fujita's work has performed a valuable service in classifying and comparing tornadoes. Remarkably, Fujita completed his scale without ever seeing a tornado. According to *The Weather Book*, written by Jack Williams, "Even though he began studying tornadoes soon after coming to the United States in 1953, and was called 'Mr. Tornado' in a 1972 *National Geographic* magazine article, Fujita didn't see a tornado until June 12, 1982, 'a date I'll remember after I forget my birthday.'"

motorists and mobile home residents to seek shelter in a building or, if one is unavailable, lie low in a field or an empty area.

Some regions of the United States are more likely to experience tornadoes than others. This is particularly true for an area of the Midwest known as Tornado Alley.

TORNADO ALLEY

Tornadoes can strike anywhere in the United States (including Hawaii and Alaska), but two regions are particularly susceptible. These two regions are the Southeast and a broad swath through the middle of the country known as Tornado Alley, as illustrated in the following figure.

Parts of the Southeast, including Louisiana, Mississippi, Alabama, and Florida, experience a lot of thunderstorms as warm, moist air from the Gulf of Mexico mixes with cold air from the north. In addition, hurricanes and other marine storms often come ashore at these states. These storms lose their strength as they travel inland, but their winds often create tornadoes in the process.

Tornado Alley is an especially prolific area for intense tornadoes at the upper end of the enhanced Fujita scale. Many of the strongest storms in the United States occur in this region.

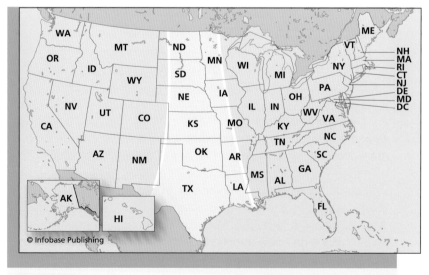

Tornado Alley

Unlike hurricanes, there is no tornado "season." Hurricanes need warm water to serve as an energy source and are limited to the summer and early autumn months in the Atlantic Ocean (see chapter 5). Tornadoes can happen at any time. There are, however, certain periods of the year in which tornadoes are more likely. These periods are different for the different areas of the country. In the Southeast, tornadoes tend to occur more frequently in the late winter and early spring. The southern and middle portions of Tornado Alley get a lot of tornadoes in the spring, while in the northern section, along with parts of the southwestern United States, tornadoes peak in the summer.

Why do so many strong tornadoes form in Tornado Alley? The simplest answer to this question is that conditions in this part of the country, along with the Southeast, are more favorable for tornado formation. As discussed in the following section, strong tornadoes often develop from certain types of thunderstorms. An area that has a lot of thunderstorms can expect some number of them to generate tornadoes.

The simplest answer is correct as far as it goes, but it is not entirely satisfying. Until researchers fully understand how tornadoes form, no one will know exactly why so many thunderstorms spawn powerful tornadoes in Tornado Alley. One possible explanation involves the movement of air masses and the specific geography of the Great Plains, which covers much of Tornado Alley. Cold, dry air flows out of the West, across the Rocky Mountains and into the plains. Warm, humid air arrives from the South, having picked up moisture from the Gulf of Mexico. In the spring, large temperature and pressure differences in these air masses create winds—wind is the flow of air from high pressure to low pressure. The overlying air masses generate *wind shear,* which refers to a situation in which wind speed or direction changes across a given area (for example, winds may increase in speed and shift in direction with increasing altitude in a certain column of air). These winds may begin to circulate, making tornadogenesis more likely.

To gain a better understanding of the conditions in which tornadoes form, researchers study the weather systems that tend to develop them. These systems include a specific type of thunderstorm called a *supercell.*

SUPERCELLS ARE DANGEROUS SYSTEMS

Thunderstorms are sometimes called "convective storms" because of the importance of convection—the transfer of heat by a flowing fluid

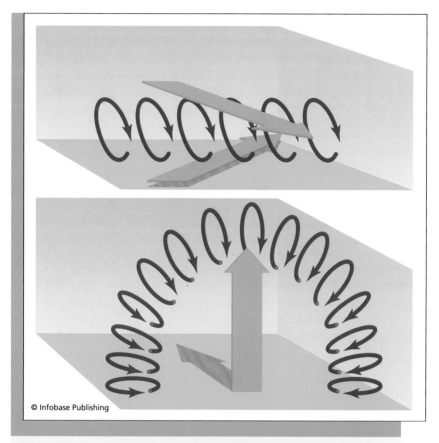

The upper part of the figure shows a rotation caused by wind shear; the bottom part shows an updraft lifting the spinning air, resulting in the creation of a mesocyclone.

such as air. Warm air rises because it is less dense than colder air. As the warm air rises it expands and cools, and the water vapor it carries begins to condense, forming water droplets. Condensation occurs because the amount of water vapor that air can carry depends on the temperature—warmer air carries more water vapor—and as the temperature falls, the vapor begins to condense around particles called condensation nuclei (see chapter 6). Clouds form in this manner.

For a thunderstorm to develop, the air needs to rise even farther. Updrafts can occur because of an especially hot day or the arrival of a cold air mass that pushes up against the air above it. A gain in altitude cools the air even more, causing more condensation and the start of downdrafts—falling air (and water droplets). At this stage, the system contains both updrafts and downdrafts. A tall, anvil-shaped cloud has formed, with heavy rain, wind, and lightning bolts (which consist of the flow of electricity that occurs because the violent motion of the storm has separated positive and negative electric charges).

There are several types of thunderstorm. Each updraft is known as a cell, and small, simple thunderstorms consist of a single cell. Other systems consist of multiple cells and are called multicell storms. The cells may form a cluster, or they may form a straight line, known as a *squall line*. Multicell clusters are the most common thunderstorms.

Supercell over Miami, Texas *(NOAA Photo Library, NOAA Central Library; OAR/ERL/ National Severe Storms Laboratory [NSSL])*

In some cases, a thunderstorm may become a supercell. A supercell is a large, well-organized, and often long-lasting thunderstorm that contains a rotating column of air—a mesocyclone. As is the case with all thunderstorms, supercells need a supply of warm, rising air, and at least one updraft. In addition, supercells require something to get the rotation started. As shown in the figure on page 90, wind shear may provide the necessary impetus. Suppose the wind at high altitude has a greater speed than at low altitude. This situation imparts a spinning motion to the air in the middle altitudes. If an updraft lifts this spinning air, some of the circulation may become vertical.

Mesocyclones are extremely important because they become part of a tornado if they extend all the way to the ground. Weather forecasters and watchers are eager to detect the presence of a supercell and its mesocyclone. To observe a mesocyclone, meteorologists often use a specific type of radar called Doppler radar.

Storm Chasers

In the 1950s, when weather radar coverage was much spottier than today, meteorologists had trouble tracking storms. To help locate and study tornadoes, adventuresome scientists such as the meteorologist Neil Ward (1914–72) began making field observations. Ward and other researchers correlated visual observations near the storm with what Doppler radar operators at the weather stations were seeing, which helped meteorologists identify radar signatures of the storms. These field researchers drove to the site of storms and followed them, as best they could, using whatever roads or pathways were available. They became known as tornado chasers or, as is more often the case, storm chasers, since many of the systems tracked fail to produce tornadoes.

Although storm chasers' early purpose was to supplement weather station observations and help police track the storms, their efforts evolved into a more active research role in the 1970s and '80s. The development of laptop computers and portable Doppler radar sets, along with improved wireless communications and tracking equipment such as the global positioning system (GPS), help storm chasers make accurate measurements in the field. Storm chasers' goals today include taking precision readings in and around storms as they develop, or fail to develop, into tornadoes. Meteorologists use the data to help understand how storms evolve and dissipate.

Getting close to tornadoes is not for the faint of heart, but scientific storm chasers are well trained and careful. *The Weather Book* quotes the University of Oklahoma storm chaser Howard Bluestein on this subject: "We believe we

Doppler radar uses the Doppler effect, named after the Austrian physicist Christian Doppler (1803–53). In 1842, Doppler discovered that the observed frequency of a wave depends on the relative motion

Storm chasers studying a developing storm in Oklahoma *(Jim Reed/Photo Researchers, Inc.)*

understand the structure of storms and we do not put ourselves in any danger of being hit by a tornado. The biggest danger is from driving on narrow country roads. The other danger is lightning."

As one of the most thrilling kinds of scientific research, storm chasing has been depicted in films such as *Twister,* released in 1996. But *Twister* was not an accurate portrayal of storms or storm chasers—which is not the first time that movie producers have sacrificed scientific accuracy for the sake of a dramatic plot or spectacular special effects. More realistic depictions include the Discovery Channel series *Storm Chasers.*

of observer and source. If an object emits a sound wave or a light wave while moving away from an observer, the frequency of the wave, as measured by the observer, is lower. The reason for the decrease is that as the object moves away from the observer, the waves spread out, increasing the wavelength and decreasing the frequency. When the motion is toward the observer, the frequency increases. This effect is noticeable when a moving train blows its whistle—as it nears a stationary observer, the pitch gets higher as the frequency increases, and then the opposite occurs after the train passes the observer and begins to move away. Astronomers use the Doppler effect to calculate the motions of a star or galaxy by studying frequency shifts of its spectrum.

In meteorology, Doppler radar emits a beam of low-frequency radiation such as radio waves or microwaves. Radar works by detecting the reflection of this radiation coming from objects such as airplanes. Weather radar uses the signals to detect the presence of water droplets or ice crystals. In addition to detecting the presence of precipitation, Doppler radar measures the shift in frequency of the reflection. If the object that reflected the radio waves was moving, then the Doppler effect applies to the reflected waves in the same way that it does to an emission of waves. With Doppler radar, meteorologists can detect movement such as the circulation of a mesocyclone.

Mesocyclones are important, but not all tornadoes form the same way. Landspouts, for example, do not originate in supercells. Although these tornadoes are interesting and can be hazardous, the remainder of this chapter will focus on tornadoes that arise in severe thunderstorms—this class of tornado generally produces the most intense winds.

Strong tornadoes occasionally do not appear to have been preceded by well-organized mesocyclones. And not all supercells generate tornadoes. Spotting a mesocyclone is not difficult with the aid of Doppler radar, but researchers have yet to determine how such rotations make the transition to tornadoes. This issue is at the forefront of much tornado research.

TORNADOGENESIS—BIRTH OF A TWISTER

Finding a mesocyclone might be easy, but detecting a tornado from a distance is not that simple, for it is often hard to tell if the circulating winds have touched the surface. Even detecting a tornado from the ground can be difficult if the funnel is hidden in the dust and debris

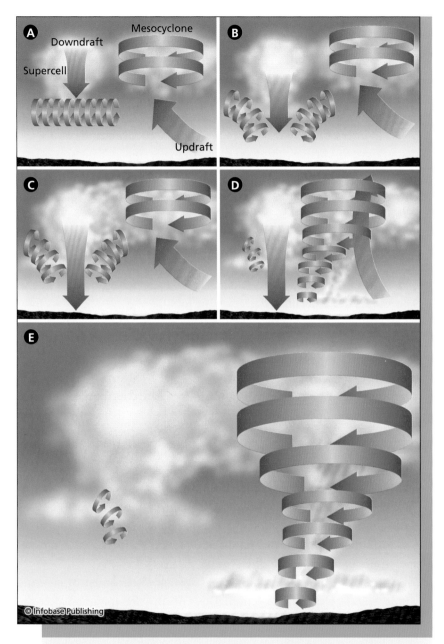

Hypothetical stages of tornado formation: (A) horizontal rotation gets caught in a downdraft; (B) two nearly vertical rotations result; (C) one of the rotations nears the mesocyclone; (D) the rotation joins the mesocyclone; (E) a tornado results.

raised by the strong winds. In order to confirm that a tornado occurred, the NWS usually investigates the damage. A survey of the damage will help weather officials determine the direction and strength of the winds, which will indicate the intensity as given by the enhanced Fujita scale.

Studying sporadic phenomena such as tornadoes is a challenging task. To gather data, researchers ideally want to know in advance where the object of interest will be and what time it will occur. Tornado researchers do not have this luxury, although they can make educated guesses—Tornado Alley is a productive place, especially in the spring. Yet there is a lot of ground to cover. Fixed weather stations can sometimes yield valuable data if they happen to be in the vicinity of a tornado's path, but most tornado researchers need to be mobile. Getting a closer look means following the track of severe thunderstorms and waiting for a tornado that may or may not develop. People who engage in this kind of research are known as storm chasers or tornado chasers. The sidebar on pages 92 and 93 provides more details on this exciting job.

Weather stations across the country employ Doppler radar, and many television stations also own and operate this type of equipment for their meteorology reports. Researchers are using information from these observations, along with data obtained by storm chasers, in the attempt to determine exactly how tornadoes form. A number of different hypotheses have emerged, but scientists do not presently agree on which one is correct.

One hypothesis involves the connection of two rotating columns. As illustrated in the figure on page 95, the rotating air generated by wind shear might drift into a downdraft. Similar to what happens when an updraft creates a mesocyclone, the downdraft pushes the rotating air down, deflecting and tilting it at an angle. By some poorly understood process, the spinning air that has been tilted joins with the mesocyclone and stretches all the way to the ground.

Large systems such as hurricanes, which also consist of swirling winds, spin in a certain direction. The Coriolis effect, named after the French scientist Gustave-Gaspard Coriolis (1792–1843), is an important factor. Coriolis showed that the rotation of Earth beneath moving objects causes a deflection in their path as they go from one latitude to another. This effect is negligible except over broad areas—either the object must travel a great distance or it must be spread out over a large area. Forces arising from the low air pressure at the center of the hurricane are also critical (see chapter 5). As a result, hurricanes rotate counterclockwise (as viewed from above) in the Northern Hemisphere and, because the effect switches directions in the Southern Hemisphere, clockwise in that hemisphere.

Most tornadoes in the Northern Hemisphere also rotate counter-clockwise—but some rotate the other way. Tornadoes are much smaller than hurricanes, which can extend for hundreds of miles. Because of their smaller size, tornadoes do not experience as great a deflection, and the motion and development of tornadoes are extremely complicated. Although the Coriolis effect may impart some tendency to spin in a cer-tain direction, as suggested by the predominance of the counterclock-wise direction in the Northern Hemisphere, the existence of clockwise-spinning tornadoes demonstrates that other factors that have not yet been clearly defined are also contributing.

Some researchers studying tornadogenesis have been focusing their efforts on developing mobile instruments of increasing quickness and precision. In 2005, the National Science Foundation (NSF) announced a mobile Doppler radar set called Rapid-Scan Doppler on Wheels (DOW) that will provide the most comprehensive view of tornadoes and how they form. (NSF, a government agency devoted to supporting basic scientific research, provided the funds for this project.) Early versions of DOW have been deployed since 1995, but the new equipment has been significantly enhanced. A press release issued by NSF on June 1, 2005, described the changes: "Most Doppler radars transmit only a single beam, which takes about five minutes to make the vertical and horizontal scans needed for a three-dimensional storm portrait. But tornadoes can develop or dissipate in a minute or less. With its five- to 10-second resolution, the Rapid-Scan DOW can detail these critical steps in tornado behavior at close range."

DOW and other well-equipped vehicles will participate in tornado projects. One of the most important experiments in recent times was called VORTEX, which stands for Verification of the Origins of Rota-tion in Tornadoes Experiment.

VORTEX: VERIFICATION OF THE ORIGINS OF ROTATION IN TORNADOES EXPERIMENT

In 1994 and 1995, about 100 researchers from the United States and Can-ada, led by Erik Rasmussen, collaborated on VORTEX. The researchers

National Severe Storms Laboratory (NSSL)

Established in 1964, the NSSL collects and analyzes storm data in order to gain a better understanding of how storms evolve, as well as improving the accuracy of weather warnings and forecasts. NSSL, located in Norman, Oklahoma, is a continuation of the efforts of the National Severe Storms Project, which had been based in Kansas City, Missouri. In its early years, one of the main goals was to optimize the use of radar for the monitoring of potentially dangerous storms.

Today, NSSL employs about 135 people and consists of three research divisions—forecast, warning, and radar. The laboratory also partners on many projects with the Cooperative Institute for Mesoscale Meteorological Studies at the University of Oklahoma (the main campus of which is also located in Norman), along with the U.S. Navy, Air Force, and Army, the Department of Transportation, the Federal Aviation Administration, and others.

NSSL has made many contributions to the study of severe weather. Researchers at NSSL refined Doppler radar techniques and constructed the first real-time display of velocity information (which means the instrument displays the measurements as they are made rather than recording and replaying them later), and acquired the first direct record-

focused their efforts on Texas, Oklahoma, and Kansas—in the heart of Tornado Alley—chasing dozens of storms, some of which generated tornadoes and some of which did not. Teams of investigators manned 12 instrumented vehicles, a mobile Doppler radar, and two airplanes equipped with Doppler radar. In order to obtain as much data as possible on individual storms, VORTEX researchers concentrated their attention on only one storm at a time rather than simultaneously chasing a number of de-

Severe storms often produce dangerous lightning strikes. *(NOAA Photo Library, NOAA Central Library; OAR/ERL/National Severe Storms Laboratory [NSSL])*

ing of a tornado made with an airborne Doppler. Current projects include participation in the Weather Research and Forecast Model, which aims to improve warning systems, and the development of polarimetric radars, which use polarized (oriented) radio waves to boost scan rates.

veloping systems. Researchers conducted operations during 18 days in the spring of each of the two years of the project.

The National Severe Storms Laboratory (NSSL), located in Norman, Oklahoma, sponsored the project along with the University of Oklahoma's Center for the Analysis and Prediction of Storms. NSSL's forecasting reports informed researchers where the best opportunities lay, and its "nowcasting" helped VORTEX teams position themselves in order to collect data safely. NSSL is a branch of NOAA and plays an important role in the study of dangerous storm systems such as tornadoes. The sidebar on pages 98 and 99 provides more information on this laboratory and its mission.

The data collected during VORTEX gave scientists a close-up view of supercells and tornadoes. Although there was not enough data for researchers to determine everything that happens during the evolution of a tornado, scientists made some important discoveries. For instance, one of the tornadoes, which appeared near Dimmitt, Texas, developed without a large temperature gradient near the surface. A gradient is a variation or change, and the absence of a significant temperature change is surprising considering the importance of the meeting of warm and cool air masses for thunderstorm formation. The Dimmitt tornado along with a tornado studied near Newcastle, Texas, formed within about half an hour, much faster than researchers were anticipating. In general, VORTEX data showed that the generation of tornadoes occurs in shorter time intervals and in smaller areas than previously believed.

VORTEX researchers also performed detailed wind measurements and were able to correlate the wind readings and tornado damage with great precision. These results contributed to the development of the enhanced Fujita scale. The data also helped to improve the accuracy of the NWS's tornado warning operations.

But there is much more to learn. Tornado researchers still need to know more about how tornadoes form and why some of these storms are powerful and long-lasting while others fizzle out in a few minutes. To make additional progress, a second round of VORTEX, called VORTEX2, began in May 2009. This two-year project, led by the National Center for Atmospheric Research (NCAR) scientist Roger Wakimoto, includes researchers from NOAA, NCAR, 10 universities, as well as participants from Canada and Australia. The target area is again the Great Plains of the United States.

VORTEX2 has an impressive armada of vehicles consisting of 10 mobile radars, including the DOW, along with advanced radars from the University of Oklahoma, NSSL, the University of Massachusetts, Texas Tech University, and the Office of Naval Research. The technological improvements pioneered by NSSL and elsewhere will be of great service, enabling VORTEX2 to perform many real-time measurements. Research teams will roam Tornado Alley for several weeks in the spring of 2009 and 2010, responding quickly to the development of severe weather in their area. With such a considerable array of equipment trained on a storm system, researchers will be able to collect massive amounts of valuable data, although the 2009 effort, which lasted five weeks, was not as productive as researchers had hoped because of a relative scarcity of tornadoes.

One of the obstacles researchers face in the study of these complex weather systems is that they are not all alike. For example, the Purdue University researcher Robert Trapp and his colleagues recently examined data from more than 3,800 tornadoes. A summary of the results of this research, which was funded by the NSF, appeared in an NSF news release issued on November 8, 2005. In the report, Trapp noted, "In the heart of Tornado Alley, twisters most often develop from relatively small 'cell' storms that look like blotches on a Doppler radar weather map. Over time, these cells frequently merge into line-shaped storms that can stretch hundreds of miles. The conventional wisdom has been that line storms don't often spawn tornadoes, but we found that a significant number did."

Trapp and his colleagues found that 79 percent of their tornado sample developed from cells (including supercells, which are "relatively small" compared to line storms), while 18 percent rose from squall lines. In the upper Midwest, however, there was a higher percentage of tornadoes from squall lines; in Indiana, for example, about 50 percent of tornadoes evolved from line storms. "This implies that we may be overlooking many tornado-breeding storms in the Midwest and elsewhere," warned Trapp. "The upshot of our analysis is that tornadoes form under a broader set of circumstances than meteorologists once thought, and this is especially true if you live far from Tornado Alley."

DEATH OF A TORNADO

Variation in tornadoes and how they develop means that researchers must broaden their scope in order to achieve a comprehensive understanding of these systems. Another important aspect of tornadoes must also receive consideration—the final stage of its evolution, when the tornado dissipates. The formation of a tornado is a crucial step in its evolution, but it is only the first stage. How long a tornado lasts is a critical factor in the length of its path and how much destruction it leaves behind.

The funnel of small tornadoes often shrinks as it dissipates. In the process, the funnel tilts and may adopt a bent or deformed shape. The last phase is sometimes called the rope stage, as the funnel resembles a rope or string. (A tornado can still be dangerous at this time.) Larger tornadoes often disappear as the funnel seems to rise into a low-lying cloud.

As is the case for the formation of tornadoes, scientists are unsure of the details of how these storms die. Storms need some kind of energy supply to sustain them, such as the updrafts of thunderstorms or warm water in the case of hurricanes. Roger Edwards, a meteorologist at the Storm Prediction Center (which is part of the NWS), noted in his Online Tornado FAQ that there are a number of phenomena that could reduce a tornado's lifetime. "One is relatively cold outflow—the flow of wind out of the precipitation area of a shower or thunderstorm. Many tornadoes have been observed to go away soon after being hit by outflow. For decades, storm observers have documented the death of numerous tornadoes when their parent circulations (mesocyclones) weaken after they become wrapped in outflow air—either from the same thunderstorm or a different one." But Edwards also mentioned that in some cases, outflows can help create the conditions that lead to tornado formation.

UNUSUAL TORNADOES

There is a wide variety in tornadoes, which includes some that behave in a strikingly different manner than most. Every tornado probably has some peculiar or unique aspects, although most of the storms follow similar patterns or stages. Every once in a while, an odd or remarkable tornado develops. For instance, the Tri-State Tornado of 1925 lasted an astonishing three and one-half hours and cruised through 219 miles (350 km). Such a long-lived twister is extremely rare.

Some people have noted that few tornadoes seem to strike heavily populated and highly developed urban areas, which is fortunate because a major tornado, such as one in the EF4 or EF5 category, could cause thousands of casualties and billions of dollars in damage if it struck a major metropolis. Perhaps the conditions leading to tornado formation are not as prevalent around big cities. But the rarity of big city tornadoes may instead be due to chance—large cities, even sprawling ones, do not cover a high percentage of any given state's area. Although a number of tornadoes develop each year, any given small area, such as a particular field or a city, has a low probability of getting hit. But people are more likely to notice when big cities are spared, rather than thinly populated sections of the country, which may account for the conception that urban tornadoes are rarer than rural ones.

On several occasions, tornadoes have struck major cities in the United States. One of the most recent was an EF2 in Atlanta, Georgia, on March 14, 2008, the first recorded tornado in the city's history. This moderate-strength tornado struck at 9:40 P.M., 10 minutes after the NWS issued a tornado warning, according to an article posted on March 15, 2008, at Foxnews.com. The winds damaged skyscrapers, blowing out some of the windows and leaving shards of glass scattered along sidewalks, and buffeted several sports arenas, which were filled with fans watching basketball games. No one was killed, but officials reported more than two dozen people suffered injuries.

Researchers at Purdue University and the University of Georgia decided to look for any special circumstances that may have contributed to the development of this tornado. The scientists, led by the Purdue professor Dev Niyogi, obtained data from the NASA satellite *Aqua* as well as the *Tropical Rainfall Measuring Mission* satellite to study precipitation levels prior to the storm and the condition of vegetation in the area. Around the time of the tornado, Atlanta and the surrounding region had been suffering from a drought, although periods of scattered rain fell in parts of the area a few days before the tornado.

Niyogi and his colleagues proposed that the scattered rainfall produced wet areas surrounded by dry ones. Air rich in water vapor may have lingered over the wet areas, setting up pockets of humid air. The air over the city may have also been considerably warmed by what is called the urban heat island effect—cities are generally hotter than the surrounding areas because asphalt and construction materials absorb and store more heat than vegetation. Wet and dry masses of air may have set up mild *fronts*—boundaries between two air masses having different densities, which are the sites of much meteorological activity, such as storms. If Niyogi and his colleagues are correct, the urban landscape, along with an uneven distribution of rain, intensified the storm on March 14, perhaps contributing to the formation of the tornado. But much more research is needed before these conclusions are confirmed.

TORNADO WARNINGS AND FORECASTING

According to the American Meteorological Society (AMS), about 55 Americans die each year from tornadoes. The storms also cause about $400 million worth of damage per year on average. Researchers hope that an increased understanding of how tornadoes form and dissipate will help reduce the number of casualties and help citizens become better prepared.

Forecasting has already improved significantly in the last few decades. The NWS is responsible for monitoring severe weather and issuing storm watches and warnings. Meteorologists there issue about 1,000 watches and 30,000 warnings for severe storms per year. (About 3,000 of these warnings are for tornadoes.) Warnings indicate a storm has been spotted or, in the opinion of the NWS, is imminent. Prior to the 1990s, the lead time—the amount of time between the warning of an imminent tornado and the actual event—was about five or six minutes. In recent times, the NWS says that the average has increased to about 12 to 15 minutes.

Tornado damage *(viZualStudio/Shutterstock)*

Although 12 to 15 minutes do not seem like a lot of time, it is often sufficient for people in the path of a storm to find shelter. The number of casualties has fallen to about half the rate before the 1990s. Additional warning time also gives endangered citizens a chance to secure their property, decreasing the extent of damage.

Meteorologists credit several events for the increased lead times. Data from VORTEX and other storm-chasing projects have enabled meteorologists to pick out tornado threats more clearly from their radar scans. The NWS's installation of a network of Doppler radars in the 1990s also advanced the quantity and quality of information available on developing storms across the country.

Tornado warnings can be improved even further. About one-third of the tornadoes that occur in the United States arrive unannounced—no warning had been issued. And sometimes warnings are issued for tornadoes that fail to show up; although people are relieved to escape unscathed, these false alarms do not encourage confidence in the warning system. Too many false alarms reduce the effectiveness of the system because weary citizens will tend to ignore future warnings.

VORTEX2 researchers expect that the data they collect will enhance and improve the NWS's warnings. With their armada of mobile technology, VORTEX2 teams should be able to refine radar signatures of tornadoes, increasing scientific knowledge of tornado formation as well as reducing the number of tornadoes that develop without warning, and cutting down on the number of false alarms. This data may also increase the lead time and provide ample warning for most tornadoes.

Researchers are also studying modes of forecasting over the longer term. Weather forecasting is exceptionally difficult more than a day or two in advance because of the complex systems and interactions involved. Severe storms are particularly complicated, making forecasting an even more formidable task.

Meteorologists at the Storm Prediction Center perform this difficult service as best they can. In his Online Tornado FAQ, Roger Edwards wrote, "When predicting severe weather (including tornadoes) a day or two in advance, we look for the development of temperature and wind flow patterns in the atmosphere which can cause enough *moisture, instability, lift,* and *wind shear* [italics in original] for tornadic thunderstorms." But those factors alone are not sufficient to make a forecast. "'How much is enough' of those is not a hard fast number, but varies a lot from situation to situation—and sometimes is unknown! A large variety of weather patterns can lead to tornadoes; and often, similar patterns may produce no severe weather at all. To further complicate it, the various computer models we use days in advance can have major biases and flaws when the forecaster tries to interpret them on the scale of thunderstorms."

Tornado forecasting makes heavy use of models, as do many scientists who study weather and climate. Given the current conditions, models predict the evolution of the system based on the laws of physics and chemistry and include statistics on the outcomes that meteorologists have observed in the past when confronted with similar conditions. But as Edwards notes, "Real-time weather observations—from satellites, weather stations, balloon packages, airplanes, wind profilers and radar-derived winds—become more and more critical the sooner the thunderstorms are expected; and the models become less important."

The additional information gleaned from VORTEX2 and other tornado research projects will help. Tornado forecasting is unlikely to ever become precise because of the complexity of Earth's atmospheric phenomena—the same factors that generally limit the accuracy of weather forecasts, which often prove wrong if they are made more than a few days in advance. But as researchers gain more knowledge of tornadogenesis, tornado forecasting will undoubtedly improve.

CONCLUSION

Models that help researchers analyze complex phenomena such as tornadoes are often computer programs. The program incorporates equations and formulas that govern the interactions of important variables such as wind speed, wind shear, pressure, and temperature. A computer user inputs some key data or variables, and the program does the rest, calculating the theoretical evolution of the system. Researchers can fine-tune the model by comparing its predictions with the actual results, as measured by Doppler radar and storm chasers.

Another way to bring the study of tornadoes into the laboratory is to mimic their properties with special equipment. Engineers often build miniature versions of airplanes or structures such as buildings and bridges in order to test their responses in certain situations and gauge their likely performance. For example, researchers subject models of jet fighters to wind tunnel tests to analyze the design's aerodynamics and air flow.

Simple tornado generators create a vortex with fans positioned in or around a chamber. To visualize the rotation, operators introduce smoke or mist into the chamber. These tornado generators are popular science fair projects and are fun to watch. But miniature versions of tornadoes that simulate actual storms are extremely difficult to set up in the laboratory. Researchers are striving to build more elaborate laboratory devices to help them study tornado properties and mechanisms.

Another important goal for this research is to study the ways in which tornadoes damage buildings and how architects and construction engineers can design structures to withstand severe winds. Scientists can study rotating winds and the damage they cause in simulators by creating a relatively simple (though large) vortex and placing models of various structures in the chamber. One of the largest simulators is at the Wind Simulation and Testing Laboratory at the College of Engineering at Iowa State University. The device is called a tornado/microburst simulator. (A microburst is a kind of downdraft—a column of sinking air—that produces dangerous winds.) The chamber is 20 feet (6.1 m) wide, 44 feet (13.4 m) long, and 18 feet (5.5 m) high. A fan with a 6-foot (1.8-m) diameter creates a vortex with a diameter of about 3 feet (0.9 m).

Partha P. Sarkar, the director of the Wind Simulation and Testing Laboratory, recently used the simulator to study how people can design structures to be more resistant to a tornado's rotating winds. In an article posted on April 22, 2008, at Iowa State Daily.com, Sarkar said, "We have learned that winds of an F2, which is a moderate tornado, will be large enough to rip roofs off of houses, buildings and other structures, so we have to make the roof load 1.5 times larger. We have to improve the design so it can withstand a moderate tornado." Since moderate tornadoes are far more frequent than those at the highest end of the scale, following these guidelines could result in substantial reductions in future tornado damage.

But simulators and models will not replace storm chasers and Doppler radar anytime soon. Tornadoes are extremely complex systems. Although scientists have made significant progress in understanding how some kinds of tornado form, the details remain sketchy. Researchers at the frontier of science must continue to tackle this puzzle—sometimes venturing out into the rain and wind with their maps, laptops, and mobile Doppler radars in search of yet more valuable data.

CHRONOLOGY

1680   First recorded fatality from a tornado in the United States occurs at Cambridge, Massachusetts.

1841   The American meteorologist James P. Espy (1785–1860) publishes *The Philosophy of Storms,* which explains the convective theory of thunderstorms.

1880s   The American meteorologist John P. Finley (1854–1943) pioneers the scientific study of tornadoes.

1925   The Tri-State Tornado devastates parts of Missouri, Illinois, and Indiana.

1950s   The American meteorologist Neil Ward (1914–72) and others begin chasing storms in order to collect data.

1964   The National Severe Storms Laboratory is founded.

NOAA commissions the first Doppler radar for weather.

1971   The Japanese physicist Tetsuya Theodore "Ted" Fujita (1920–98) proposes a scale to gauge tornado intensity.

1974   A "super outbreak" of 148 tornadoes strikes parts of the South and Midwest regions of the United States on April 3–4.

1980s   Storm chasers unsuccessfully deploy the Totable Tornado Observatory (TOTO) in the path of tornadoes in the attempt to collect data from the inside. Safety concerns and technical problems force these efforts to conclude in 1987.

1994–95   Verification of the Origins of Rotation in Tornadoes Experiment (VORTEX) takes place in Tornado Alley.

1995	The first Doppler on Wheels hits the road.
1999	Doppler on Wheels records a wind speed of 318 miles/hour (509 km/hr) during an Oklahoma tornado.
2005	The Rapid-Scan Doppler on Wheels debuts.
2007	Meteorologists adopt the enhanced Fujita scale.
2008	An EF2 tornado sweeps through downtown Atlanta, Georgia.
2009–10	The second Verification of the Origins of Rotation in Tornadoes Experiment (VORTEX2) takes place. Well-equipped storm chasers roam the Great Plains in search of storms to track and study.

FURTHER RESOURCES
Print and Internet

Baker, Tim. "Tornado Alley: Tornado and Storm Chaser Facts, Pictures, and Weather Information." Available online. URL: http://www. tornadochaser.net/. Accessed July 1, 2009. The storm chaser Tim Baker maintains this Web resource devoted to the art and science of tracking tornadoes.

Bluestein, Howard B. *Tornado Alley: Monster Storms of the Great Plains.* Oxford: Oxford University Press, 2006. Bluestein, a meteorologist at the University of Oklahoma, has spent many years studying and chasing tornadoes. This book discusses the origins and progress of tornado research, including some of Bluestein's many adventures in the field.

Edwards, Roger. "The Online Tornado FAQ." Available online. URL: http://www.spc.noaa.gov/faq/tornado/. Accessed July 1, 2009. Edwards, a meteorologist at the Storm Prediction Center, compiled and answered this excellent list of frequently asked questions about tornadoes.

Foxnews.com. "Tornado Rips through Downtown Atlanta, Injuring at Least 27" (3/15/08). Available online. URL: http://www.foxnews.

com/story/0,2933,338078,00.html. Accessed July 1, 2009. This report describes the EF2 tornado that struck Atlanta, Georgia, on March 14, 2008.

Grazulis, Thomas P. *The Tornado: Nature's Ultimate Windstorm.* Norman: University of Oklahoma Press, 2003. Grazulis, a meteorologist, describes a tornado's formation and life cycle, winds and wind speeds, the Fujita scale, tornado forecasting, and tornado myths and safety.

Iowa State Daily.com. "ISU Tornado Simulator Drew National Attention" (4/22/08). Available online. URL: http://www.iowastatedaily. com/articles/2008/04/22/fyi/20080422-archive0.txt. Accessed July 1, 2009. Iowa State University's tornado simulator has received widespread coverage in the news media.

Mathis, Nancy. *Storm Warning: The Story of a Killer Tornado.* New York: Touchstone, 2007. Mathis, a journalist from Oklahoma, recounts the remarkable series of tornadoes that scourged Oklahoma on May 3, 1999.

National Aeronautics and Space Administration. "Drought, Urbanization Were Ingredients for Atlanta's Perfect Storm." News release (3/11/09). Available online. URL: http://www.nasa.gov/topics/earth/ features/atlanta_tornado_prt.htm. Accessed July 1, 2009. Researchers at Purdue University and the University of Georgia used NASA satellite data to study the conditions leading up to the tornado in Atlanta, Georgia, on March 14, 2008.

National Oceanic and Atmospheric Administration. "Tornadoes . . . Nature's Most Violent Storms." Available online. URL: http://www. nssl.noaa.gov/edu/safety/tornadoguide.html. Accessed July 1, 2009. This guide explains the basics of tornadoes and what to do if a storm is approaching.

National Science Foundation. "Rapid-Scanning Doppler on Wheels Keeps Pace with Twisters." News release (6/1/05). Available online. URL: http://www.nsf.gov/news/news_summ.jsp?cntn_id=104209. Accessed July 1, 2009. An NSF-funded project has installed a rapid-scan Doppler radar on a vehicle in order to provide high-resolution scans of tornadoes.

———. "Scientists Unravel Midwest Tornado Formation." News release (11/8/05). Available online. URL: http://www.nsf.gov/news/

news_summ.jsp?cntn_id=103157. Accessed July 1, 2009. The Purdue University researcher Robert Trapp and his colleagues discover considerable variation in the evolution of tornadoes.

Williams, Jack. *The Weather Book,* 2nd ed. New York: Vintage Books, 1997. Williams, one of the founding editors of the weather section in the newspaper *USA Today,* describes and explains weather systems with the aid of many colorful illustrations. Chapter topics include tornadoes, floods, snow and ice, hurricanes, and others.

Web Sites

Center for Severe Weather Research. Available online. URL: http://www.cswr.org/. Accessed July 1, 2009. The Center for Severe Weather Research is a nonprofit research organization that participates in much tornado research, including the Doppler on Wheels project. Their Web site offers data archives and information on Doppler on Wheels and other research programs.

National Severe Storms Laboratory. Available online. URL: http://www.nssl.noaa.gov/. Accessed July 1, 2009. NSSL's Web site provides news and information on their latest research efforts.

Storm Prediction Center: Monthly and Annual United States Tornado Summaries. Available online. URL: http://www.spc.ncep.noaa.gov/climo/online/monthly/newm.html. Accessed July 1, 2009. This Web site provides statistics on tornadoes by year and region.

VORTEX: Unraveling the Secrets. Available online. URL: http://www.nssl.noaa.gov/noaastory/book.html. Accessed July 1, 2009. The Verification of the Origins of Rotation in Tornadoes Experiment, conducted in 1994 and 1995, was one of the largest tornado hunts in history. This Web site describes the project.

VORTEX2. Available online. URL: http://www.vortex2.org/home/. Accessed July 1, 2009. This Web site provides news and information on the second VORTEX project, which began in May 2009.

HURRICANE FORECASTING

On September 8, 1900, the deadliest natural disaster in U.S. history occurred when a hurricane lashed Galveston, Texas. Located on Galveston Island, the city was a densely populated port and one of Texas's richest cities. At this time, hurricanes were not named, but the 1900 Galveston storm was a memorable one. High winds and floodwaters swept over the city, killing at least 6,000 people and probably a few thousand more—the final tally will never be known because of the devastation and chaos that followed, and many of the bodies washed out to sea.

Galveston had not been fully prepared. Satellites did not exist in 1900, and although residents knew a storm was churning through the Gulf of Mexico—meteorologists had received reports from ships and islands—no one knew the strength of the storm or the track it would take.

But hurricanes are such powerful storms that disaster can strike even when cities have been warned. Hurricane Katrina came ashore on the southern coast of the United States early in the morning of August 29, 2005, killing about 1,800 people and doing tens of billions of dollars worth of damage. The storm hit New Orleans, Louisiana, especially hard; located between Lake Pontchartrain and the Mississippi River, most of New Orleans lies below sea level, protected by embankments called levees. Hurricane Katrina breached some of these levees, flooding about 80 percent of the city and creating more destruction than had been anticipated.

There has been much hurricane activity recently, and 2008 was a particularly active year with five major storms. Meteorologists monitor these storms with extensive networks of weather stations, reports from airplanes and ships, and satellite imagery, making it impossible for a hurricane to sneak up on anybody today. But scientists do not yet fully understand these storms. Forecasters do not know exactly where on the coast a given hurricane will strike, so costly preparations and evacuations must cover a broad area. False alarms encourage a potentially deadly complacency among coastal residents. Predicting the storm's intensity at landfall is also difficult; some storms weaken just before they strike—to the surprise and relief of those in the strike zone—but some storms mysteriously strengthen, turning a mild storm into a killer.

Researchers have made progress recently in forecasting hurricanes, but puzzles remain. Intrepid researchers fly into hurricanes in order to collect valuable information, and other researchers analyze reams of data from all over the globe searching for clues about hurricane development and the paths these storms take. This chapter describes the current state of research and what remains to be accomplished. A more comprehensive view of these powerful storms will lead to improved forecasts as well as a better understanding of Earth and its weather and climate.

This image of New Orleans, taken by satellite on September 6, 2005, shows a portion of the Mississippi River running across the center and Lake Pontchartrain at the top. In between, most of the city of New Orleans is flooded. *(NASA image courtesy Lawrence Ong, EO-1 Mission Science Office, NASA GSFC)*

INTRODUCTION

The launch of the Russian satellite *Sputnik* in 1957 ushered in the space age, and soon thereafter meteorologists were using orbiting instruments to study weather and climate. Early researchers did not have such high vantage points. Marine storms were poorly understood until American saddle-maker and self-taught scientist William Redfield (1789–1857) studied the aftermath of a storm that came ashore in Connecticut in 1821. Redfield noticed that felled trees pointed toward the northwest in one part of the damaged region, indicating that they had been blown over by a southeast wind, but the remains of trees in another area pointed toward the southeast, indicating a northwest wind! After further investigation, Redfield proposed that these storms consist of rotating winds—whirlwinds.

Satellite images show that hurricanes are large marine storms with winds circling around a central eye. The term *hurricane* derives from the Spanish word for hurricane, *huracán,* which early Spanish explorers in the New World adapted from a similar word that Caribbean inhabitants used. A hurricane's eye is a calm region of exceptionally low pressure, with high-speed winds swirling around it. A satellite image usually displays a small, circular eye in the middle of a patch of clouds that extends 300 miles (500 km) on average, but can reach 500 miles (800 km) or more for gigantic storms. Radar images show that hurricanes consist of bands of thunderstorms revolving around the eye. Winds are strongest near the *eyewall.* Hurricane winds can reach speeds of about 200 MPH (320 km/hr); for a storm to be classified as a hurricane, sustained winds must reach a speed of at least 74 MPH (119 km/hr). A sustained wind is one that maintains its speed for a long period, unlike a gust of wind, which lasts only a few seconds.

Although the whirling wind of hurricanes is similar to tornadoes, the two kinds of storm are quite different. Tornadoes are smaller and can have higher wind speeds, but tend to last only a few minutes or a few hours at most, while hurricanes are large and can live for more than a week.

The mechanics of a hurricane are also different because they form over water. A hurricane is a kind of heat engine—it converts the energy of warm water into wind. Hurricanes require ocean temperatures of at least 80°F (26.7°C) and a lot of moist air. As described in the following section, the storms evolve from weather disturbances and begin to rotate due to the Coriolis effect. A hurricane rapidly loses wind speed and falls apart over land.

The need for warm water means that hurricanes tend to form only during certain months. In the Atlantic, hurricane season begins on

June 1, when the Atlantic Ocean is generally warm enough to permit hurricane formation, and ends November 30. Almost all hurricanes in the Atlantic Ocean arise during this six-month period, with the peak coming around August or September, although hurricanes will occasionally occur before June or after November. The 2008 hurricane season set a record in that five of the six months of the season experienced a major hurricane; the old record was 2005, when four of the six months had a major hurricane. Meteorologists consider a hurricane "major" if it is in Category 3 or higher, as described below. In the last five decades, an average of about six hurricanes have formed in the Atlantic Ocean every year, though the number

Satellite image of Katrina in the Gulf of Mexico *(NASA/ Jeff Schmaltz, MODIS Land Rapid Response Team)*

ber varies considerably from year to year. Some turn north and die in the cold North Atlantic, while some go on to hit land—about two on average strike the U.S. mainland each year.

A hurricane is a strong *tropical cyclone*—a storm characterized by winds whirling around an area of low air pressure that forms over tropical waters. The term *hurricane* tends to be used in the United States instead of tropical cyclone, while *typhoon* is the term commonly used for strong Pacific Ocean cyclones. Most tropical waters—warm waters in the Tropics, around the equator—experience cyclones, except for the South Atlantic and the southeastern Pacific Ocean, where the wind pattern tends to preclude their development. Hurricanes that strike the United States almost always form in the Atlantic, where they move in an easterly direction and hit the East Coast or the coast along the Gulf of Mexico. Pacific Ocean hurricanes tend to move westward, away from the West Coast of the United States. A few Pacific hurricanes have struck the western coast of Mexico, but there is no record of any hitting the United States.

Saffir-Simpson Scale—Hurricane Categories

Meteorologists in the 1960s and early 1970s were having trouble communicating the strength and potential danger of hurricanes to officials who were responsible for disaster preparedness. The old saying that a picture is worth a thousand words can also apply to a simple number, which is easy to state and can convey a great deal of information. At about the same time that the Japanese physicist Tetsuya Theodore Fujita (1920–98) developed the Fujita scale for tornadoes, an American engineer Herbert Saffir (1917–2007) and the meteorologist Robert Simpson (1912–) developed a five-category scale for ranking hurricanes.

Saffir created the initial version of the scale in 1971 when he described the expected damage that hurricanes of specific intensities may cause. Simpson added wind speed and flooding potential, creating the Saffir-Simpson scale. Because flooding has proven to be so variable, only the wind speed and damage are generally used today. The five categories are as follows:

Meteorologists classify hurricanes according to sustained wind speed and the potential damage a storm may cause. The classification scheme consists of five categories, with Category 1 being the weakest and Category 5 the strongest. The sidebar above provides more details.

Storms are also given names, a practice that began in 1953. The World Meteorological Organization (WMO) maintains six lists of names that are used in rotation—one list is used one year, then the next list, and so on, cycling back to the first list after the sixth year. Atlantic and Pacific Ocean storms have different lists. The names are alphabetical, one for each letter (except for Q, U, X, Y, and Z in the Atlantic list).

- Category 1: Minimal damage, wind speed of 74–95 MPH (119–153 km/hr)
- Category 2: Moderate damage, wind speed of 96–110 MPH (154–177 km/hr)
- Category 3: Extensive damage, wind speed of 111–130 MPH (178-209 km/hr)
- Category 4: Extreme damage, wind speed of 131–155 MPH (210–249 km/hr)
- Category 5: Catastrophic damage, wind speed greater than 155 MPH (249 km/hr)

Although meteorologists began using the scale in the early 1970s, many storms that occurred before this time have been categorized based on damage reports and meteorological data.

Category 5 storms are rare, occurring only once every few years on average, and few storms are this strong when they make landfall—most weaken considerably before they reach the shore. Only three recorded hurricanes have hit the United States coastline as Category 5 storms: an unnamed hurricane that hit Florida in 1935, Hurricane Camille, which struck Mississippi in 1969, and Hurricane Andrew, which struck Florida in 1992.

For example, the first storm in the Atlantic Ocean in 2009 was named Ana, the second Bill, the third Claudette, and so on, down to Wanda. In 2010, the first three names on the list are Alex, Bonnie, and Colin. If more storms occur in a year than names on the list, meteorologists begin using Greek letters—alpha, beta, and so on. Notable storms have their names retired; for instance, there will be no more hurricanes named Katrina or Camille. Some scientists are of the opinion that personalizing storms detracts from the seriousness of the science of meteorology and also attaches an unpleasant connotation to the names of deadly storms. But the custom is firmly entrenched and will probably continue for some time to come.

BIRTH AND DEATH OF A HURRICANE

Hurricanes develop from smaller storms that arise in tropical waters. The initial phase is a strong thunderstorm or cluster of thunderstorms. (See chapter 4 for a discussion of thunderstorms.) If these thunderstorms grow stronger, they may organize into what meteorologists call a tropical depression. A tropical depression has a region of relatively low air pressure—a "low"—with winds having a speed of 23 to 38 MPH (37–61 km/hr) circulating around the center. If the wind intensifies to a speed of 39 MPH (62 km/hr), meteorologists call the system a tropical storm and give it a name from the list. The first tropical storm in 2010, for example, will be Tropical Storm Alex.

Tropical storms may last for days, moving across the ocean with a push from upper level winds. A tropical storm might reach shore, in which case residents should be prepared for high winds, or it might fizzle out after a while. Another possibility is that the storm intensifies. If the sustained winds reach a speed of 74 MPH (119 km/hr), meteorologists refer to the system as a hurricane. (Why 74 MPH [119 km/hr] and not some other speed? This speed has traditionally been used as the minimum for the severest storms, as calculated by the British admiral Sir Francis Beaufort [1774–1857] in 1805.) If Tropical Storm Alex intensifies to this extent, it will get upgraded to Hurricane Alex.

Scientists are not sure exactly how tropical depressions form, or why some of these storms intensify and progress to the hurricane stage. Observations and data accumulated over the years suggest the following scenario. Water from the warm ocean evaporates and rises. The rising air creates an area in which the air density is decreased—an area of low pressure. Low pressure draws air from surrounding regions as the atmosphere attempts to equalize the pressure; high pressure air naturally flows toward regions of low pressure until equilibrium is restored. This creates wind. As the moist, rising air encounters a significantly cooler temperature high above the surface, it begins to condense. Condensation releases heat, which warms the air and causes it to rise to greater altitudes. Winds begin to blow more strongly.

The energy from condensation is an important component in the "heat engine" of hurricanes, as illustrated in the figure at right. This energy is called latent heat because it is stored or hidden in the phase of a

Warm, moist air rises, and the water vapor condenses as the temperature cools. Condensation releases heat and warms the air, which begins to rise higher.

substance. For example, heat melts ice because it agitates the molecules, which breaks the bonds holding the solid together. Heat also boils water by breaking the weak bonds holding the molecules in the liquid phase.

National Hurricane Center

In 1870, President Ulysses Grant signed a congressional resolution forming the Weather Bureau (forerunner of the NWS). Its mission was to make meteorological observations and warn communities about the approach of severe weather, including storms that threatened the coast. In 1898, President William McKinley went further, ordering the establishment of a hurricane warning network. These early efforts did not have the advantage of airplane reconnaissance and satellite imagery and failed to provide sufficiently accurate warning—the 1900 Galveston hurricane, for example, became the nation's deadliest—but the success rate increased along with technology. In 1965, an office of the NWS in Miami, Florida, became the NHC.

In its early years, the NHC focused on the Atlantic Ocean. The center's meteorologists monitored the formation of storms in the Atlantic, tracking their paths and forecasting where the storms appeared to be headed. When meteorologists believe that a hurricane may strike a given area within 36 hours, they issue a hurricane watch for that area. If meteorologists anticipate a hurricane will strike within 24 hours, they issue a hurricane warning. The NHC is now also responsible for monitoring the eastern Pacific Ocean as well.

Former NHC directors include Robert Simpson (cocreator of the Saffir-Simpson scale), director from 1967 to 1973, and Neil Frank, from 1973 to 1987. Frank was often called "Mr. Hurricane" because of his passion for the job and his many television appearances, especially when the center was tracking a major storm that threatened the United States. The current director is William L. Read, a meteorologist whose experience includes gathering data by flying into hurricanes.

This energy is stored rather than lost because it gets released when the gaseous molecules condense into liquid.

As the updraft intensifies, the pressure in the center of the storm falls. The winds pick up speed. But the winds rotate rather than blowing in a straight line because of the Coriolis effect. The Coriolis effect, sometimes called a "force," is a deflection due to Earth's rotation. Different latitudes on Earth's surface turn at different speeds as the planet spins on its axis (a point at the equator, for example, moves the entire circumference of Earth in a single day, while a point near one of the poles hardly moves at all). As an object moves long distances, it appears to curve with respect to the surface because of the variation in speed with latitude. The Coriolis effect deflects winds toward the right in the Northern Hemisphere and toward the left in the Southern Hemisphere. But the pressure difference is paramount. Winds of a tropical cyclone go swirling around the central eye—the "low"—trying to reach the low pressure but not quite getting there. In the Northern Hemisphere, cyclones rotate counterclockwise (as viewed from above).

A division of the National Weather Service (NWS) called the National Hurricane Center (NHC) is responsible for monitoring hurricane activity and issues hurricane watches and warnings. The sidebar on page 120 provides more information about this important agency.

Some hurricanes lose strength and fall apart over cooler water before they reach land, but other hurricanes make landfall. Damaging winds that topple trees and buildings can hit a coastal community six or more hours before the hurricane's eye arrives. Storm surges are also extremely dangerous. A storm surge consists of waves of water pushed by the wind; water has a lot of mass, and the crashing waves behave like battering rams, crushing anything in their path. Category 3 hurricanes typically have a storm surge of about nine to 12 feet (2.7–3.7 m). However, Camille, a 1969 Category 5 hurricane that hit Mississippi, had a storm surge of 24 feet (7.3 m), and Hurricane Katrina's storm surge exceeded 30 feet in parts of the Gulf Coast.

Hurricanes weaken and dissipate once they move over land. But some storms continue to be dangerous; Camille and Katrina, for example, maintained hurricane force winds—74 MPH (119 km/hr)—for about 100 miles (160 km) inland.

Meteorologists at the NHC need a great deal of data in a timely manner in order to monitor the world's oceans. Methods of data acquisition include satellites, radar, reports from ships and passenger airliners, and meteorologists who are willing to fly in and around storms.

HURRICANE HUNTERS

Hurricane hunters make direct meteorological measurements by flying into the teeth of the storm. Before the development of weather satellites, aviators provided spotting and tracking services, as well as collecting important data. These days, satellites can easily monitor the oceans and track developing storms, but satellite instruments are remote and limited in the kind of measurement they can make. Satellites cannot generally penetrate the cloud cover to see inside the storm or take accurate barometric (pressure) and wind speed readings. The job of collecting this data falls to the men and women who are willing to endure a bumpy ride for a look at a hurricane's interior.

A WC-130J Hercules hurricane hunter (U.S. Air Force photo/Tech. Sgt. James Pritchett)

Several organizations currently fly "hurricane hunting" missions. The U.S. Air Force Reserve's 53rd Weather Reconnaissance Squadron, based at Keesler Air Force Base in Biloxi, Mississippi, has a complement of 10 WC-130J Hercules aircraft. The squadron's primary job is tropical storm reconnaissance over a broad area, including the Gulf of Mexico, the Atlantic Ocean up to 55°W longitude, and the Pacific Ocean up to the international date line. Ten full-time crews man the airplanes, along with 10 part-time crews for active periods—the squadron is capable of covering three storms at a time. In the hurricane off-season, the crews collect data on winter storms.

Most of the missions of the 53rd Weather Reconnaissance Squadron are classified as operational meteorology, meaning that the crews are usually helping meteorologists track and forecast a specific storm. Other hurricane hunters come from the National Oceanic and Atmospheric Administration (NOAA), which often performs more research-related missions. Although NOAA hurricane hunters also participate in helping meteorologists forecast a particular storm, their missions include taking readings to help researchers understand how storms form and evolve.

NOAA uses two different kinds of aircraft. One, the WP-3D Orion, flies through the hurricane. P-3D is a turboprop—propeller-driven, like the WC-130J, sturdy and maneuverable at low altitudes in hurricane weather. But for fast speeds and high-altitude flights, such as soaring over the top of a hurricane, NOAA employs a Gulfstream IV SP jet. These airplanes are maintained at NOAA's Aircraft Operations Center at MacDill Air Force Base in Tampa, Florida.

Hurricane hunters deploy instruments called dropwindsondes, or just sondes, which derives from a French word for sounding line. These instruments have global positioning system (GPS) to track their location. As they fall, they measure variables such as air pressure, humidity, wind speed and direction, and temperature and transmit the data back to the aircraft.

The missions are dangerous but essential for the study, tracking, and accurate forecasting of hurricanes. And the ride is not bumpy all the time—the worst part is the strong wind near the eye. Once inside the calm eye, hurricane hunters have spectacular views of the storm. Only one airplane has been lost flying into an Atlantic hurricane—a Navy P-2, lost in 1955 on a mission to Hurricane Janet. Three planes have crashed while flying into Pacific Ocean storms.

TRACKING A HURRICANE

The NHC, aided by satellites, radar, and hurricane hunters, monitors and tracks storms. Hurricanes usually do not stay still over one spot in the ocean. Although occasionally a storm will stall and spin around a stationary point for some time, most hurricanes are on the move.

What causes a hurricane to move? The main driving force consists of upper-level winds at an altitude of about 12,000 to 40,000 feet (3,700–12,200 m). Meteorologists often refer to these winds as steering currents.

Winds arise because of differences in the pressure of air (air pressure is also called barometric pressure because of the use of instruments known as barometers to measure it). Air pressure decreases with increasing altitude, but it also varies from point to point at a given altitude. These variations are small but important, because they are the reason for winds—air flows from high to low pressure. (Gravity balances the vertical air pressure differences, so they do not generally create wind.) Meteorologists note areas of high (H) and low (L) pressure on maps. The pressure is measured relative to the surrounding region; a high on a map is at higher pressure than the air around it, and a low is at lower pressure. Heat and temperature differentials create these kinds of pressure variations. For example, sunshine heats the ground on a hot summer day, which increases the temperature of the surface as well as the air immediately above it; this air rises, creating an area of low pressure. The rising air drifts and then cools, falling elsewhere, creating a region with increased density—an area of high pressure.

Trade winds are global wind patterns that arise in regions with consistent pressures. High-pressure belts girdle the globe at around 30° latitude (in both the Northern and Southern Hemispheres) as cool air masses from the poles sink. Winds blow from this region toward a steady low pressure around the equator and are curved by the Coriolis effect. The result is a fairly consistent wind from the northeast blowing toward the southwest in the Northern Hemisphere in the lower latitudes. These winds helped early sailors on their voyages from Europe to America (for the return trip, the ships took a more northern route, where the prevailing winds are easterly).

Westerly winds push Atlantic hurricanes toward the American coast. Tropical storms often form off the western coast of Africa and begin moving across the Atlantic. Meteorologists sometimes call these systems Cape Verde storms, after the Cape Verde Islands, located about 400 miles (640 km) off the coast of West Africa, which are in the vicinity of many developing storms. These storms usually have the best chance of becoming major hurricanes because of the plentiful amount of warm water in front of them. But other storms, including some that grow quite strong, originate in the Caribbean.

Tracking hurricanes becomes even more critical as they near the shore, threatening coastal communities. To increase the accuracy of their equipment, NHC began implementing a new system in 2008 called

Vortex Objective Radar Tracking and Circulation (VORTRAC). This system uses information from Doppler radars arrayed on the U.S. coast to determine an approaching hurricane's vital statistics—wind and central barometric pressure. In a press release issued on April 10, 2008, by the National Science Foundation (NSF), which funded the development of this system, a NHC meteorologist Colin McAdie said, "VORTRAC allows us to take the wind measurements from the radar, turn the crank, and have a central pressure drop out of a calculation. This will be a valuable addition to the tools available to the forecaster."

FORECASTING THE PATH OF THE STORM

Meteorologists give special attention to hurricanes that appear headed for landfall instead of turning northward and dissipating in the cold North Atlantic Ocean. Although tracking the storms as they approach land is important, meteorologists must keep a close watch on storms that are well out to sea. Coastal residents need to be warned as quickly as possible.

Major hurricanes destroy homes and buildings with wind and storm surges, so officials order residents in the affected area to evacuate. But evacuating a large metropolitan area such as Miami or Houston takes at least a few days—the roads can hold only so much traffic, and they get jammed when everyone starts to leave at the same time. Officials in coastal towns and cities prepare for these emergencies, but orderly evacuations take some time. Late warnings will not enable all residents to leave, and, what is worse, the storm might catch vulnerable motorists stuck in traffic jams.

One solution to this problem would be to warn coastal communities any time a hurricane poses a risk of landfall. But with so many storms arising in the Atlantic, citizens would be packing up numerous times each hurricane season. Evacuations are costly to a region's economy as businesses close and people temporarily leave their jobs, so unnecessary evacuations should be avoided. And a succession of false alarms would probably induce apathy or mistrust, which might result in a dangerous lack of preparation when a storm finally arrives.

To get an idea where a storm might be heading while it is still in the middle of the ocean, meteorologists analyze pressure zones and upper-level winds—the steering currents. Hurricane forecasters measure these

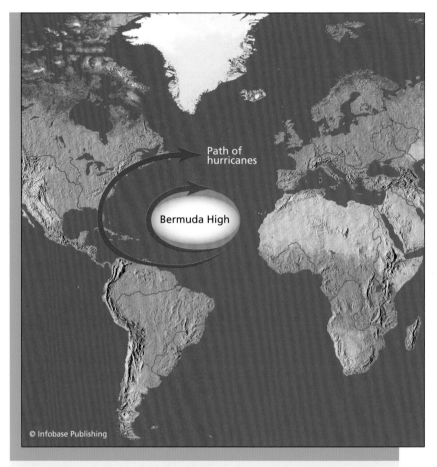

Hurricanes swerve around high-pressure zones. If the high does not cover much area then it does not push the hurricane very close to the U.S. coast, but if the high has expanded it steers the hurricane toward America and the Caribbean islands.

winds by releasing instrumented weather balloons. Weather stations send these balloons aloft twice a day. Additional information comes from NOAA airplanes, as well as commercial airline pilots, who often report upper-level winds during their flights.

Another important factor guiding Atlantic hurricanes is the Bermuda High, a region of high pressure that tends to form in summer months over the Atlantic Ocean near the island of Bermuda. The extent of this high-pressure region influences the path of storms, as illustrated in the figure above. When the Bermuda High covers a large area, its high pressure pushes storms toward the coast of the United States.

Steering currents, pressure zones, and how they interact with specific hurricanes and their properties are extremely complicated. Meteorologists use sophisticated computer models to make hurricane forecasts. These models take data collected at weather stations and by hurricane hunters and churn through calculations in order to predict the future course of a storm. Two main types of models are statistical and dynamical. Statistical models focus on the behavior that previous storms exhibited under similar conditions; the idea is that the course of a certain storm is statistically likely to follow the same path as a similar storm that developed under the same conditions. Dynamical models employ equations of physics that govern the behavior of the atmosphere and phenomena such as hurricanes. All of these models have strengths and weaknesses, depending on the circumstances. According

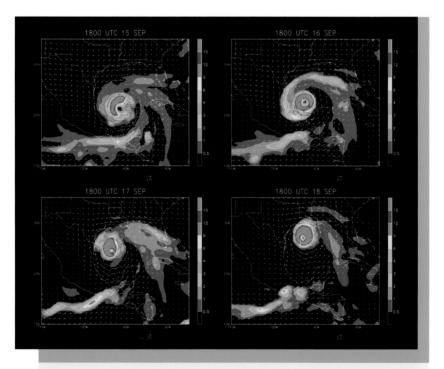

Computer simulation of the development of a hurricane, with the colors representing precipitation—red is the highest amount of precipitation and blue is the least *(Science Source)*

to the meteorologist Jamie Rhome at the NHC, "A given NHC forecast never relies solely on any one individual model (i.e., 'model of the day' or 'best model'), but rather reflects consideration of all available guidance as well as forecaster experience."

With these tools, meteorologists narrow the range of possible landing sites. But uncertainty remains. As with all weather forecasts, the uncertainty increases the farther out that forecasters try to see—a forecast one day in advance is more accurate than one four days in advance. This introduces a cone of uncertainty, as the predicted path of a storm becomes increasingly uncertain—and therefore broader in area—the farther removed it is from present time. According to NHC statistics covering 2003–07, the average error in landfall predictions for Atlantic hurricanes during this period was 272.5 nautical miles (505 km) when the storm was 120 hours from making landfall, 106.2 nautical miles (197 km) when the storm was 48 hours from making landfall, and 34 nautical miles (63 km) when the storm was only 12 hours away from the coast. These errors have been steadily decreasing for many years. For instance, the 48-hour error in the early 1970s was about 300 nautical miles (556 km), almost three times higher than in 2003–07.

But when no specific steering current predominates, hurricanes tend to meander, making forecasting even more difficult than usual. As meteorologists gain more experience and the models improve as hurricane hunters and other researchers accumulate more and richer data sets, forecasts will achieve even greater precision. New techniques such as VORTRAC should also reduce the size of the uncertainty. Even though the 48-hour forecasts are accurate enough to prevent many unnecessary evacuations, researchers aim to do better in the future.

The intensity of the storm is as important as its track. Knowing whether a hurricane will be weak or strong at the time of landfall is vital for preparation, otherwise officials may evacuate too many people or not enough. But forecasting a hurricane's intensity has been a bigger problem for meteorologists than predicting the path.

PREDICTING STORM INTENSITY

Many hurricanes gain strength slowly, peaking at a Category 2 or 3, and then holding steady or fading slightly until dissipating over land or cool water. But there are important and dangerous exceptions. Hurricane

Andrew, which devastated southern Florida in 1992, was intensifying as it came ashore. In 1969, Hurricane Camille strengthened from a Category 1 to a Category 5 in just two days—in 48 hours the storm went from a minimal threat to a killer storm. Researchers still do not understand how Camille strengthened so quickly.

A strong hurricane is well organized, with winds whipping around a distinct eye that shows up clearly in satellite images. Storms that are weakening become disorganized as the winds die down or scatter in different directions and the eye disappears. The size of the eye is also a measure of intensity, since smaller diameters are associated with stronger winds; the average eye diameter is about 20 to 30 miles (32–48 km).

The Florida International University researcher Hugh E. Willoughby noted in a 2007 issue of *Science*, "Track forecasts have improved steadily over the years, but intensity predictions have lagged a generation behind track forecasts." One of the problems is that models that forecast intensity "need finer grid resolution to represent the small-scale, rapidly evolving organization of convection in the hurricane core. For this reason, improvements of intensity forecasts have been slower."

To improve the models, researchers are looking for factors associated with changes in hurricane strength. The Florida State University researchers Xiaolei Zou and Yonghui Wu recently used a National Aeronautics and Space Administration (NASA) satellite with an ozone-mapping spectrometer to study ozone levels in 12 hurricanes. Ozone consists of three atoms of oxygen (O_3), which exists in low concentrations in the atmosphere. As announced in a NASA news release issued on June 8, 2005, the researchers found that a certain area of a hurricane "typically has low levels of ozone from the surface to the top of the hurricane. Whenever a hurricane intensifies, it appears that the ozone levels throughout the storm decrease." Although the reason for this drop is not clear, the correlation may become useful in forecasting models.

Other researchers are focusing on the eyewall—the strong thunderstorms and high-speed winds that surround the eye. Many storms proceed through cycles as they develop. Occasionally during a storm's lifetime a new batch of thunderstorms crops up close to but at some distance from the eyewall. These thunderstorms rob some of the warm, humid air that fuels the strong winds of the eyewall, so they begin to slow down and the eye widens. Soon these new thunderstorms replace the bands of old thunderstorms that comprised the eyewall. The storm is losing strength.

But after eyewall replacement, a storm may intensify as it settles down to business again and the fuel begins to flow freely. The eye shrinks as the winds pick up speed. (The expansion and shrinkage of the eye with wind speed is due to the conservation of angular momentum. Angular momentum is the product of an object's size and its rate of rotation—for a given amount of angular momentum, a smaller size results in a higher rotation rate. A similar effect can be seen with a spinning ice-skater—the rate of spin increases if the skater retracts his or her arms and decreases with the arms extended.)

In 2005, Robert Houze of the University of Washington, Shuyi Chen of the University of Miami, and a team of researchers at a number of institutions, including NOAA, conducted a series of investigations known as the Hurricane Rainband and Intensity Change Experiment (RAINEX). Sponsored by the NSF, RAINEX used the two NOAA P-3 hurricane hunters, along with a Navy P-3, to study the activity around a hurricane's eyewall. The year 2005 was a fortuitous choice for the hurricane researchers, as several major hurricanes occurred. RAINEX obtained a great deal of data from Hurricanes Katrina, Ophelia, and Rita.

The researchers focused their efforts on eyewall replacement. Their hypothesis was that the changes in intensity taking place in this process involved certain interactions in the bands of rain and storms. RAINEX used high-resolution airborne Doppler radar and coordinated the readings with satellite data. Advanced models and "nowcasts" guided the researchers as they steered their planes through the hurricane, searching for regions having the potential for the most activity.

Houze, Chen, and their colleagues published some of the findings of this study in a 2007 issue of *Science*. The researchers noted significant strength reductions during eyewall replacement. In their paper, they wrote that their data "suggests that with a sufficiently high-resolution model, the eyewall replacement and accompanying intensity decrease can be forecast. In addition, the successful documentation of the eyewall replacement process in Rita during RAINEX by ground-controlled aircraft targeting of the small-scale features internal to the tropical cyclone vortex could be a harbinger of improved forecasting of hurricane intensity. The adoption of real-time targeting of aircraft onto small-scale storm features likely to be associated with storm intensity change could provide timely input that would improve operational forecasts of hurricane intensity."

SEASONAL PREDICTIONS

Every form of weather forecasting, including hurricane forecasts, suffers from the complexity of the systems that meteorologists are trying to predict. The Massachusetts Institute of Technology meteorologist Edward Lorenz (1917–2008) discovered in the 1960s that the predictions of weather models diverged widely even with tiny changes in the relevant variables. This effect is called chaos theory, or simply chaos, and is a mathematical property of certain complex systems such as those involved in weather. These systems are highly sensitive to changes in conditions, and since complicated conditions are impossible to specify exactly, the system's behavior is virtually impossible to predict, at least over the long term. Lorenz described this extreme sensitivity as the *butterfly effect*—a butterfly flapping its wings in Brazil might lead to a change in conditions that could touch off a tornado in the United States.

Sensitivity to conditions is the reason why the accuracy of hurricane forecasts decreases the more they are made in advance. Long-term forecasts and seasonal predictions—how many storms meteorologists expect for a given hurricane season—are especially challenging. But trends and patterns have begun to emerge.

Strong hurricanes have tended to come in cycles recently. More than a dozen major hurricanes hit the United States from 1941 through 1965, but only four struck from 1966 to 1994. The frequency began to rise again in 1995. Meteorologists must determine if this cyclical activity is meaningful or is just due to chance.

William Gray, a researcher at Colorado State University, is one of the pioneers of seasonal hurricane forecasts. He and other researchers noticed that an important factor involved a phenomenon known as El Niño—a periodic warming in temperature of the surface of the Pacific Ocean in the eastern equatorial region. Fishermen in Peru have long noted these episodes, which occur every two to seven years, because they affect fishing stocks and agriculture. Because it usually begins around Christmas, it is called El Niño, Spanish for the boy, referring to the birth of Christ. As discussed in the following sidebar, El Niño and its opposite, La Niña (Spanish for little girl), are part of an important

El Niño, La Niña, and the Southern Oscillation

An episode of El Niño begins when temperatures in the tropical waters of the eastern Pacific Ocean begin to rise. This is one stage of a cycle in which the temperature of these waters rises and then falls. An average El Niño lasts about a year. La Niña refers to a period, typically lasting a few years, in which the temperature is cooler than usual. These two phases cycle back and forth. Large-scale variations in air pressure over these waters accompany this cycle. The air pressure fluctuations are known as the Southern Oscillation. During an El Niño episode, pressure rises over the western tropical Pacific Ocean and falls over the eastern section. The opposite situation occurs in La Niña. Together, the water and air pressure cycle is known as El Niño-Southern Oscillation (ENSO).

ENSO directly affects South Americans who live along the western coast and nearby islands. In the 1950s, meteorologists realized ENSO wielded a much broader influence

oscillation taking place in the tropical Pacific Ocean. Researchers cannot yet predict when these events will occur, but they closely monitor conditions in order to spot important changes as they develop.

Since the 1950s, scientists have noted widespread weather and climate events accompanying these periodic changes. One of the changes involves the number of hurricanes. During strong episodes of El Niño, the eastern Pacific Ocean tends to experience more typhoons and the Atlantic Ocean has fewer hurricanes. La Niña has the opposite effect.

How do El Niño and La Niña influence the formation of hurricanes? The changes in water temperature help explain why El Niño fosters cyclone development in the Pacific Ocean and La Niña discourages it, but wind patterns are also important. Adverse winds, such as vertical wind on the world's weather and climate. During the 1957–58 International Geophysical Year, in which researchers made a concentrated effort to study Earth and its environment, researchers discovered correlations between an ongoing El Niño and weather patterns around the globe. ENSO and its phases can affect conditions in places far removed from the tropical Pacific. Droughts and storms in the United States have been linked to specific ENSO events.

Meteorologists are not sure how El Niño exerts such widespread influence, but the magnitude of episodes strongly correlates with unusual weather patterns. Temperature and pressure do not change much during some ENSO cycles, but at other times the changes are much greater; for example, sea-surface temperatures (SST) can increase several degrees across vast portions of the Pacific Ocean. An El Niño occurring in 1982–83 was particularly strong, as was the 1998–99 episode. The consequences included enormous increases in stormy weather and rainfall in parts of the United States, causing serious flooding, and severe droughts in Australia and other places in the western Pacific Ocean.

shear, tend to prevent cyclones from getting organized. These wind factors are probably critical to Atlantic hurricanes. Earth's atmosphere is one giant ocean of air, and pressure changes in one area tend to get passed around, disrupting normal patterns. El Niño seems to create wind conditions around the Atlantic Ocean that are not favorable for cyclone formation, while La Niña has the opposite effect.

Gray and other researchers consult ENSO data before predicting what may unfold during a given hurricane season. But even with this information, along with other trends, seasonal predictions are difficult and often inaccurate. For example, researchers at Colorado State University predicted 15 tropical storms for 2005, but 28 storms actually formed.

CONCLUSION

Despite an elevation in hurricane activity recently and a rise in the number of people living on the coasts, the hurricane death toll in the United States has declined in the last few decades compared to the early and middle 20th century. Elaborate monitoring systems and increasingly accurate forecasts have helped meteorologists issue appropriate warnings. Enhanced techniques such as VORTRAC, along with research aimed at improving track and intensity prediction, should result in further gains. Damage losses have risen and will continue to do so—buildings cannot be evacuated—but lives will be saved. And scientists will achieve a better understanding of one of nature's most powerful phenomena.

But there is a possible crimp in this optimistic scenario. Although meteorologists are making progress in the study of hurricanes, Earth's environment and climate have begun to change. Global climate change may alter conditions enough that scientists will be forced to relearn a new set of rules.

Scientists are studying numerous potential consequences of the warming that has been occurring the last century over much of the globe. One consequence could be an increase in the number of storms as ocean water temperature rises, providing more fuel for hurricanes. Hurricane season may begin extending beyond six months.

The Florida State University researchers James B. Elsner and Thomas H. Jagger, along with James P. Kossin at the University of Wisconsin-Madison, recently analyzed satellite data from 2,097 tropical cyclones occurring from 1981 to 2006. The researchers determined the maximum winds for these storms and used statistical techniques to search for trends. Statistics is a branch of mathematics that quantifies probabilities and expected outcomes; scientists from nearly all disciplines, including meteorology, use these techniques to judge if a fluctuation or trend is a significant departure from the values expected by chance. By calculating the typical fluctuation in a particular variable over a period of time, researchers can determine if any given change is normal, and therefore probably due to chance or random "noise" in the system, or if it is so rare that some other force or phenomenon is probably acting on the system.

Elsner and his colleagues discovered that wind speed for the most powerful storms in the world increased from 140 MPH (224 km/hr) in

1981 to 156 MPH (250 km/hr) in 2006. During this period, the temperature of the tropical waters where cyclones form increased from 82.8°F (28.2°C) to 83.3°F (28.5°C). The researchers published their results in a 2008 issue of *Nature.* They wrote, "Our results are qualitatively consistent with the hypothesis that as the seas warm, the ocean has more energy to convert to tropical cyclone wind."

If the trend continues, cyclones may become increasingly powerful. But there are other factors involved, and not all scientists are convinced that global warming is playing a role. While rising ocean temperatures could feed more powerful storms, global climate change could induce other changes, such as altered wind patterns, that could act to disrupt storm formation.

The NOAA researcher Gabriel Vecchi and his colleagues are not certain that the uptrend will continue. They also argue that scientists have not yet proven that SST is the only factor in hurricane activity. Writing in a 2008 issue of *Science,* Vecchi and his colleagues noted, "The issue is not whether sea-surface temperature is a predictor of this activity but how it is a predictor. Given the evidence suggesting that relative SST controls hurricane activity, efforts to link changes in hurricane activity to absolute SST must not be based solely on statistical relationships but must also offer alternative theories and models that can be used to test the physical arguments underlying this premise. In either case, continuing to move beyond empirical statistical relationships into a fuller, dynamically based understanding of the tropical atmosphere must be of the highest priority, including assessing and improving the quality of regional SST projections in global climate models."

Scientists at the frontier of weather and climate science will continue to explore this fascinating subject. The complexity of cyclones and the factors involved in their formation are still incompletely understood. Much more research is needed before meteorologists will be comfortable make long-term predictions of these dangerous storms.

CHRONOLOGY

1805   The British admiral Sir Francis Beaufort (1774–1857) introduces the Beaufort wind scale.

1821	The American scientist William Redfield (1789–1857) notes the whirlwind nature of a storm that comes ashore in Connecticut.
1900	A strong hurricane makes landfall in Galveston, Texas, claiming at least 6,000 victims.
1935	A Category 5 (subsequently estimated) hurricane strikes Florida.
1943	Army Air Force colonel Joseph B. Duckworth (1903–64) intentionally pilots a plane into a hurricane, the first such recorded instance.
1945	Pilots of the 53rd Weather Reconnaissance Squadron become the first to fly a B-17 into a hurricane. The primary mission of the squadron quickly becomes "hurricane hunting."
1947	The United States Weather Bureau (forerunner of the National Weather Service) begins full-time hurricane warnings.
1957	The former Union of Soviet Socialist Republics (USSR) launches *Sputnik,* the first artificial satellite.
1965	The National Hurricane Center is established.
1969	Hurricane Camille, one of the most powerful storms on record, devastates the Mississippi Gulf Coast.
1971	The American engineer Herbert Saffir (1917–2007) devises an early hurricane scale, adapted by meteorologist Robert Simpson (1912–).
1973	In order to be closer to the Gulf Coast, the 53rd Weather Reconnaissance Squadron moves to Keesler Air Force Base in Mississippi.

1976	The first NOAA P-3 flies into Hurricane Bonny over the Pacific Ocean.
1980s	The Colorado State University researcher William Gray documents the influence of El Niño on hurricane frequency.
1997	NOAA begins flying a Gulfstream IV SP jet on hurricane research and surveillance missions.
2005	Hurricane Katrina floods New Orleans.
	Hurricane Rainband and Intensity Change Experiment (RAINEX), sponsored by the National Science Foundation, uses NOAA hurricane hunters to study eyewall replacement.
2008	Research on the possible effects of global climate change on hurricanes shows an upward trend in intensity.

FURTHER RESOURCES

Print and Internet

Atlantic Oceanographic and Meteorological Laboratory. "Frequently Asked Questions: Hurricanes, Typhoons, and Tropical Cyclones." Available online. URL: http://www.aoml.noaa.gov/hrd/tcfaq/tcfaqHED.html. Accessed July 1, 2009. This FAQ includes sections on basic definitions, storm names, winds, record storms, forecasting, cyclone climatology, history, and others.

Brinkley, Douglas. *The Great Deluge: Hurricane Katrina, New Orleans, and the Mississippi Gulf Coast.* New York: William Morrow, 2006. The author, a professor at Tulane University in New Orleans, provides a firsthand account of the disaster and its impact on the city's residents.

Elsner, James B., James P. Kossin, and Thomas H. Jagger. "The Increasing Intensity of the Strongest Tropical Cyclones." *Nature* 455

(9/4/08): 92–95. The researchers used satellite data from more than 2,000 cyclones from 1981 to 2006 and found that the strongest winds are increasing over this period.

Emanuel, Kerry. *Divine Wind: The History and Science of Hurricanes.* Oxford: Oxford University Press, 2005. Emanuel presents a comprehensive review of hurricanes, describing the history of disastrous storms and explaining what scientists know about hurricane formation and development.

Houze Jr., Robert A., Shuyi S. Chen, Bradley F. Smull, Wen-Chau Lee, and Michael M. Bell. "Hurricane Intensity and Eyewall Replacement." *Science* 315 (3/2/07): 1,235–1,239. The researchers describe some of the results of the Hurricane Rainband and Intensity Change Experiment (RAINEX).

National Aeronautics and Space Administration. "Ozone Levels Drop When Hurricanes Are Strengthening." News release (6/8/05). Available online. URL: http://www.nasa.gov/vision/earth/environment/ozone_drop.html. Accessed July 1, 2009. The Florida State University researchers Xiaolei Zou and Yonghui Wu find that the concentration of ozone falls as hurricanes intensify.

National Science Foundation. "Forecasters Implement New Hurricane-Tracking Technique." News release (4/10/08). Available online. URL: http://www.nsf.gov/news/news_summ.jsp?cntn_id=111407. Accessed July 1, 2009. The National Hurricane Center has begun to use Vortex Objective Radar Tracking and Circulation (VORTRAC) to improve tracking and forecasting of hurricanes as they near the shore.

Rhome, Jamie R. "Technical Summary of the National Hurricane Center Track and Intensity Models." Available online. URL: http://www.nhc.noaa.gov/pdf/model_summary_20070912.pdf. Accessed July 1, 2009. The NHC meteorologist Jamie Rhome describes the models the center uses to make forecasts.

Sheets, Bob, and Jack Williams. *Hurricane Watch: Forecasting the Deadliest Storms on Earth.* New York: Vintage, 2001. The authors provide a historical guide to the study of hurricanes, including chapters on computer models and attempts to control storms.

Vecchi, Gabriel A., Kyle L. Swanson, and Brian J. Soden. "Whither Hurricane Activity?" *Science* 322 (10/31/08): 687–689. The researchers discuss the effects of global climate change on hurricane activity.

Williams, Jack. *The Weather Book,* 2nd ed. New York: Vintage Books, 1997. Williams, one of the founding editors of the weather section in the newspaper *USA Today,* describes and explains weather systems with the aid of many colorful illustrations. Chapter topics include tornadoes, floods, snow and ice, hurricanes, and others.

Willoughby, Hugh E. "Forecasting Hurricane Intensity and Impacts." *Science* 315 (3/2/07): 1,232–1,233. Willoughby discusses some recent experiments on how hurricanes gain or lose strength.

Web Sites

Galveston Storm of 1900. Available online. URL: http://www.history. noaa.gov/stories_tales/cline2.html. Accessed July 1, 2009. NOAA sponsors this historical look at the Galveston hurricane of 1900, with photographs and reports.

Hurricane Hunters Association. Available online. URL: http://www. hurricanehunters.com/. Accessed July 1, 2009. This Web site describes the mission, planes, people, and history of the 53rd Weather Reconnaissance Squadron.

National Hurricane Center. Available online. URL: http://www.nhc. noaa.gov/. Accessed July 1, 2009. Visitors of the home page of the National Hurricane Center can find storm information, including satellite and radar data, as well as forecasts and articles discussing the science and history of hurricanes.

INTENTIONAL WEATHER MODIFICATION

The complexities of weather systems and phenomena make forecasting difficult, even with recent advances in satellites, radar technology, and computer modeling. Weather exhibits extreme sensitivity to conditions—small changes can have significant impacts, as described by the butterfly effect (see chapter 5). This means that tiny fluctuations that meteorologists fail to take into account can cause weather patterns to diverge from long-term forecasts after only a few days.

Sensitivity to conditions and the difficulties of forecasting also hinder any attempt to influence or control the weather. Small, seemingly insignificant changes can have drastic consequences, so anyone attempting to control the system must exercise a great deal of caution. Imagine an automobile with an accelerator that is sensitive to even the gentlest pressure or a bicycle with wheels that rotate at top speed with just a tap of the pedals. Maintaining proper control over such a vehicle would be a trying task.

But weather responds to changes in temperature, moisture, pressure, and so forth in understandable ways. As researchers gain more knowledge of the factors underlying weather systems, including severe weather such as tornadoes and hurricanes, meteorologists achieve a better understanding of how these systems form and evolve, even though forecasting remains difficult because of the sensitivity to conditions.

Humans have already influenced weather and climate, although not intentionally. Scientists from the Intergovernmental Panel on Climate Change (IPCC), and many other people believe that much of the blame for global warming lies with human activity. These scientists say that the increasing release of greenhouse gases (GHGs) has caused the atmosphere to trap heat.

Intentional modification of weather may be just as easy to produce—and as difficult to control. This chapter describes efforts to modify or control the weather. Most projects involve precipitation, usually eliciting snow or rain (or sometimes trying to prevent it), but people have entertained more ambitious projects, including weakening major storms. The topic of this chapter is controversial. Much of the controversy arises because of the difficulty in deciding if the modification actually worked. Although skepticism exists, the field is ripe with opportunities. The North American Interstate Weather Modification Council, an organization consisting of researchers and regulatory agencies involved in weather modification, maintains a list at their Web site of dozens of ongoing projects in 10 states in the United States and one Canadian province. This research field is at the leading edge of the frontiers of weather and climate science.

INTRODUCTION

One of the earliest attempts to influence the weather is the lightning rod. American statesman Benjamin Franklin (1706–90) had many other interests such as business and science. As described in the article "The Lightning Rod," posted at the Web site of the Franklin Institute (a museum and research center in Philadelphia, Pennsylvania, that is named in Franklin's honor), "By 1750, in addition to wanting to prove that lightning was electricity, Franklin began to think about protecting people, buildings, and other structures from lightning. This grew into his idea for the lightning rod. Franklin described an iron rod about 8 or 10 feet long that was sharpened to a point at the end. He wrote, 'the electrical fire would, I think, be drawn out of a cloud silently, before it could come near enough to strike . . .'"

Franklin went on to perform his famous experiments to prove that lightning is an electrical phenomenon by drawing sparks from a key attached to the string of a kite, which he flew during a thunderstorm. He also constructed lightning rods, which soon protected many structures in the United States and abroad. A lightning rod is a metal fixture that rises a small distance above the building. Since lightning

Rain falling from a cloudy sky *(Adisa/Shutterstock)*

tends to strike the tallest object in the vicinity, lightning rods draw a lot of strikes that might otherwise damage buildings. The metal conductor carries the electric charges harmlessly to the ground. Although lightning rods are simple, they effectively divert potentially hazardous weather phenomena.

While some weather phenomena are hazardous, others are much needed. People need rain to deliver supplies of fresh drinking water, and farmers must irrigate their crops. A lack of precipitation has caused severe hardships, leading people to yearn for rain during dry periods—and occasionally people have gotten desperate enough to try to bring it about.

Many early, prescientific attempts were based on superstitions. Some superstitions appear silly—a dance or invocation cannot cause rain to fall from the sky—but it is easy for superstitions to get started and propagate. If rain follows some kind of ceremony or ritual once or twice, especially during a particularly needy time, then people will be likely to continue the practice, even if it often seems to fail. The ritual worked once, the thinking goes, so it might work again.

A number of 19th-century efforts at weather modification involved loud sounds. Despite the advances in meteorology taking place during this century, scientific understanding was incomplete (as it remains today), and many people put forward theories that noise could generate rain.

In the early 1890s, the Midwest experienced a dry spell so severe that people clamored for the government to fix the situation. In 1891, Congress agreed to provide funds for a rainmaking project. The Department of Agriculture appointed Robert G. Dyrenforth (1844–1910), a patent lawyer and former soldier who had engineering experience, to lead the effort. Dyrenforth arrived in Texas with a significant military arsenal of cannons and bombs as well as a number of balloons. The idea was to blast the sky with the noise of shells fired from the cannons on the ground and explosive devices dropped from the balloons. According to the theory, the noise from these explosions would cause moisture to collect into drops and fall as rain.

During several trials in 1891 and 1892, Dyrenforth and his men fired shells from the cannons and dropped bombs while soaring aloft in the balloons. Most of the time, the only thing the project successfully invoked was ear-shattering noise. The noise or "concussion" theory of rain fell out of favor.

The modern era of weather modification began in 1946. Vincent Schaefer (1906–93), a self-taught inventor and scientist who worked for General Electric, performed an experiment at the General Electric Research Laboratory in Schenectady, New York. Schaefer discovered that dry ice—solid carbon dioxide—induced the formation of a huge number of crystals in air containing *supercooled water*. Supercooled water refers to a condition in which the temperature has been chilled below the freezing point, but the phase remains liquid instead of transitioning to a solid. For example, water will remain a liquid below its freezing point in the absence of a "seed" or nucleation site around which an ice crystal can begin to form—crystals need some sort of surface on which to start growing. Schaefer found that dry ice serves as an excellent site for crystal formation. When Schaefer introduced dry ice into a chamber of supercooled water vapor, a flurry of ice crystals formed.

In November 1946, Schaefer tested this concept in the field. He dropped dry ice shavings from a plane into clouds over western Massachusetts and was delighted to observe the subsequent falling of snow and ice crystals. Around the same time that Schaefer conducted his field test, the General Electric researcher Bernard Vonnegut (1914–97) made a similar discovery at the General Electric Research Laboratory. He discovered that silver iodide particles also serve as nucleation sites.

MAKING RAIN

Schaefer, Vonnegut, and the researchers who followed them were mimicking natural processes. Rain falls from clouds in which water droplets or ice crystals become too heavy to stay aloft.

Water droplets or ice crystals form around a *condensation nucleus.* Similar to the formation of crystals, water vapor needs a surface around which to condense—it cannot generally make the transition from the gas phase to the liquid phase without a condensation nucleus. In nature, various small particles such as dust motes, various aerosols, and salt molecules act as condensation nuclei.

Clouds are made of tiny water droplets and/or ice crystals. They are white because they reflect all components of the spectrum of light equally, although the underside of thick clouds looks dark because their bulk prevents much sunlight from penetrating. Although gravity constantly tugs at the tiny cloud droplets and crystals, air currents and updrafts are often strong enough to balance the force of gravity and keep these small, lightweight particles aloft. The following figure illustrates the formation of a typical cloud.

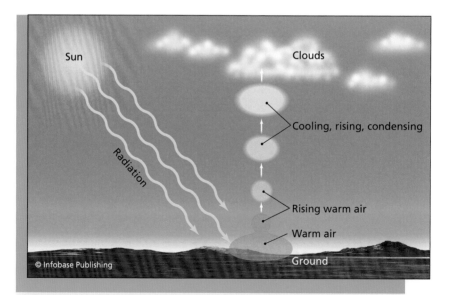

The Sun's radiation warms the ground and the air, causing the air to rise—the rising, expanding air cools and water vapor condenses, thereby forming clouds.

Precipitation begins when the particles become too heavy to be supported. Meteorologists do not fully understand how this happens, but most scientists believe that the particles grow by various means. One way is accretion, in which water droplets collide and stick. As they gain weight, the droplets start to fall, smashing into other droplets below, accelerating the increase in size. Ice crystals might grow by acquiring additional coats of water that quickly freeze to the surface. The added weight eventually brings the particles to the ground. Often the particles are frozen because of the cold temperature at high altitude, but if the air near the surface is warm enough, the crystals melt and fall as rain.

Precipitation varies throughout the year for many areas, with more falling in one season or another. But sometimes the amount of precipitation does not adhere to its usual pattern, with significantly more than the usual amount, which increases the risk of flooding, or significantly less, which results in a *drought.* As discussed in the sidebar on page 146, droughts can devastate a community.

Even under normal conditions, some regions of the world are quite dry. Little rain falls in most parts of the Sahara in Africa, for instance. Much of the western part of the United States receives as little as 15 inches (37.5 cm) on average each year (except for the coastal regions of Washington, Oregon, and much of northern California, which receive substantially more precipitation). The driest state in the country is Nevada, followed by Utah.

Sometimes the absence of rain may be due to a lack of condensation nuclei, but there must also be moisture in the air before precipitation can occur. Weather systems coming from the Gulf of Mexico pick up a lot of water vapor and bring heavy rains to the Southeast (although droughts can and do occur in this section of the country), and a similar situation occurs as storms move in from the Pacific Ocean to parts of the West Coast. But dry air hovers over much of the West, resulting in lower average participation than the rest of the country. No one can make rain fall out of dry air.

Dry conditions stifle economic development—few crops can grow, businesses are limited, and there is not enough water to support large communities. Droughts exacerbate the situation, causing dangerous shortages. In order to alleviate these problems, some communities, particularly in the West, have tried to wring out whatever amount of moisture is in the air by seeding clouds.

Droughts

Droughts are relative to the average climate of a specific region—a drought is a deficit in normal precipitation. Less than 25 inches (63 cm) of rain in a year in South Florida would constitute a serious drought in this area but would be a typical amount for parts of the Midwest. Bagdad, California, in the Mojave Desert, holds the record for the longest rainless period in the United States—767 days, from October 3, 1912, to November 8, 1914. The degree of severity of a drought depends on the rainfall deficit, how long it continues, and the water shortage that the drought produces in the affected communities.

The worst drought in American history occurred in the Great Plains states in the 1930s. Parts of Oklahoma, Texas, Kansas, and Colorado were particularly hard hit and became known as the dust bowl, named for the wind-driven dust clouds that rolled through the region. A prolonged dry spell, along with poor farming practices that left soil exposed so that it was easily dried out and blown away by winds, led to the disaster. The country's economy was mired in a depression during this time, which added to the hardships of people in this region. With no jobs, food, or water, hundreds of thousands of people packed their belongings onto trucks and moved away, resulting in the largest migration in U.S. history. Called Okies, although only a minority hailed from Oklahoma, many sought better prospects in California.

The National Integrated Drought Information System monitors conditions in the United States. At any given time, a significant portion of the country is experiencing a dry spell. For example, in May 2009, almost a third of the area of the United States suffered from rainfall deficits. Southern regions of Texas and New Mexico were in the worst shape.

CLOUD SEEDING

Experiments of Schaefer, Vonnegut, and their colleagues demonstrated how the introduction of condensation nuclei can under some circumstances increase the number of ice crystals or water droplets. Condensation nuclei in these experiments came from dry ice or silver iodide. These substances continue to be used today.

The following figure illustrates several types of cloud-seeding techniques. In some cases, airplanes release the material that will serve as condensation nuclei. Rockets fired from the ground can also deliver the material to the proper altitude. If sufficient updrafts exist, generators on the surface can do the job, releasing materials that rise into the clouds. Effective materials include dry ice, silver iodide, salt compounds, and liquid propane. Flares or liquid fuel generators can generate trillions of particles from a small amount of material.

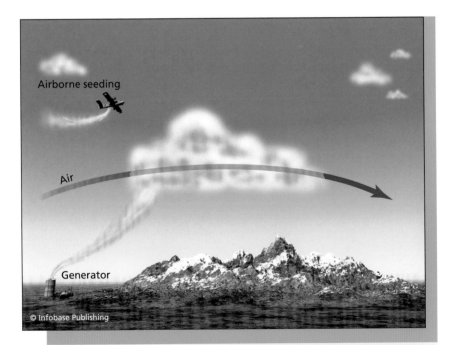

Winter clouds are seeded from the ground or the air; crystals form, some of which may fall as ice or snow after about 30 minutes.

Seeding must begin with a cloud—attempts to seed dry air will certainly fail. But not all clouds are likely candidates. In the summer months, the best candidates are clouds with updrafts of moist air that rise to altitudes high enough to freeze water, or at least contain super-cooled water. Winter clouds should also contain pockets of supercooled water. Clouds such as *cumulus clouds*—a common type with a puffy appearance (*cumulus* is Latin for "heap")—tend to be short-lived and difficult to seed. *Orographic clouds,* on the other hand, tend to be stable and long-lived, so they make better targets. Orographic clouds form when air moves over mountainous terrain and begins to rise (the term *orographic* refers to mountains and derives from the Greek word *oros,* meaning "mountain").

One of the earliest long-term projects involving weather modification was Project Cirrus. Schaefer and his colleagues at General Electric needed more resources than they had available to study rainmaking, so they turned to the government for help. (The General Electric research-ers were also leery of running into trouble if they significantly modified weather patterns without the government's knowledge or permission.) Schaefer chose the name *cirrus* because the goal of the project was to generate ice crystals such as those contained in the wispy, high-altitude clouds called *cirrus clouds.* The project began in 1947. Although the General Electric researchers participated, sponsorship and direction of the project were the responsibility of the U.S. Army Signal Corps and the Office of Naval Research, with the cooperation of the U.S. Air Force.

The Air Force provided aircraft such as B-17s to be used in the proj-ect. Early in the project, most flights took place around Schenectady, New York, the site of the General Electric Research Laboratory. But re-searchers soon realized that many opportunities to study rain and cloud seeding lay elsewhere. Project Cirrus conducted many experiments in New Mexico, home to much technical and scientific expertise because of research laboratories at Sandia and Los Alamos, as well as the site of a regular weather pattern. In the summer in central New Mexico, most days begin with a clear sky and then clouds form late in the morning over the mountains, often followed by a thunderstorm.

Because the researchers were breaking new scientific ground, the-ories and hypotheses did not drive the experiments as much as trial and error—do this, see if it works, then do something else. Project Cir-rus researchers managed to perfect silver iodide generators and other cloud-seeding mechanisms. They also studied cloud physics and gained knowledge of how precipitation occurs naturally. During cloud-seeding flights, scientists determined that the number of nuclei necessary to increase crystal production was about 10,000–50,000 per 35.3 ft^3 (1 m^3). Project Cirrus ended in 1952.

The big question was whether cloud seeding actually generated more precipitation than would have otherwise fallen. This is an extremely difficult question to answer—precipitation is relatively simple to measure, but deciding how much would have fallen without the seeding operation is tricky. Controversy arose over the results. A 1952 report about Project Cirrus published by the General Electric Research Laboratory contained the following conclusion: "The results of the various New Mexico tests, coupled with observations of the effects of other ground seeding with silver iodide, point to significant possibilities in the widespread modifying of weather conditions." But other researchers disagreed. The Project Cirrus report duly noted, "As is so often the case with the proposal of striking or revolutionary new concepts in science, the validity of the observations and conclusions of the members of the Research Group, both before and after the establishment of Project Cirrus, was challenged by many."

Controversies—and skepticism—continue. Some researchers claim to see substantial effects and others attribute any increase in precipitation to natural causes. The technology has progressed to the point where private companies perform cloud-seeding services for various government agencies. For example, officials in Utah have engaged several operators for a number of projects. The Utah Division of Water Resources has reported successful results on their Web site: "Over the years, local sponsors along with the Utah Division of Water Resources have been involved with numerous cloud-seeding programs designed to increase the winter precipitation within different areas of the state. Studies indicate that these winter seeding projects generally increase the winter precipitation by 14 to 20 percent. Economic analysis of this sort of increase in precipitation shows that the benefits from the extra water far outweigh the operational costs of seeding." The following sections discuss in more detail the issues involved with scientific evaluation of cloud-seeding projects.

Eliciting rain or snow is not the only goal in weather modification. Dispelling fog that can cause accidents and suppressing potentially damaging hail are other worthwhile projects. And sometimes people want less rain instead of more. One of those times occurred before the 2008 Olympics held in Beijing, China.

MAKING THE RAIN GO AWAY

China has invested heavily in cloud-seeding operations and technology. Part of the motivation for this investment is the same as in the western United States—farmers need rain for their crops and people need fresh drinking water. The Gobi Desert is Asia's largest and occupies a portion of Mongolia and northern China. Dry winds from this desert often sweep through northern and northwestern China, resulting in arid conditions. The Chinese try to counter this situation by coaxing rain to fall whenever they find humid air. Thousands of workers, armed with rockets and cannons to disperse condensation nuclei, conduct the cloud-seeding operations. The Beijing Weather Modification Office guides these efforts.

Chinese meteorologists use guns, such as this one near Beijing, to seed clouds.
(Diego Azubel/epa/Corbis)

Although the Chinese frequently employ cloud seeding to invoke precipitation, they are also interested in stopping rain from ruining holidays and ceremonies. The 2008 Beijing Summer Olympics was a particularly important example. Every four years, athletes from all over the world compete in Olympic contests that attract huge crowds and plenty of news coverage. Cities that host the Olympics must bid for the privilege—the International Olympic Committee selects the winners—and many countries use the opportunity to present their nation and their people in the best possible light to the rest of the world. For China, a country that sometimes engages in controversial political and social policies, the opportunity was a welcome one.

In order to keep the rain from dampening the opening ceremonies, which are attended by many world leaders, the Chinese attempted to seed clouds some distance from the site so that rain would fall elsewhere and fired rockets to disperse clouds that appeared to be moving toward Beijing. Despite the humid air and predictions for rain, the four-hour ceremony stayed dry.

Chinese meteorologists could not reproduce this success at other times during the Olympics, since wet weather forced a number of competitions to be postponed. But some rainstorms are likely to be so heavy that they cannot be averted.

But did the weather modification really work during the opening ceremonies? Some meteorologists remain skeptical, and the Chinese do not always use the most up-to-date technology. Earlier in 2008, the journalist Larry O'Hanlon talked to meteorologists about weather modification, and, while he noted some optimism, he also found a few critics. In an article posted at the Web site of the Discovery Channel, O'Hanlon wrote, "China has spent $100 million and employed 30,000 people in weather modification projects, but they are using old techniques from the 1960s and 1970s and have no way of evaluating whether their efforts are working, said Roelof Bruintjes of the National Center for Atmospheric Research, who recently visited China."

Meteorologists must determine the effectiveness of rainmaking procedures before these techniques will become widely used. Researchers at the frontier of weather and climate science are investigating this contentious problem.

EVALUATING THE EFFECTIVENESS OF RAINMAKING

Scientists agree on a number of issues. Weather can be modified—it changes frequently—and in addition to natural processes, human activity can generate the factors that cause these changes. Climate change and GHGs are one example. On the local level, urban heat islands affect the temperature and precipitation patterns around cities compared to the surrounding region. The urban heat island effect stems in part from the properties of construction materials that make up buildings—these materials tend to absorb heat, raising the temperature and affecting wind and precipitation patterns in the area.

Researchers also acknowledge that seeding causes some changes to occur in clouds. Early studies such as Project Cirrus clearly documented an increase in crystals and droplets.

But scientists do not fully understand the process, and gaps remain in their knowledge of how precipitation develops. The complexity of the interactions of air, temperature, pressure, and other factors makes atmospheric phenomena difficult to analyze. Meteorologists cannot predict the weather with complete accuracy, so they do not know for certain whether rain will fall in many situations and cannot forecast the amount of rainfall with unerring precision.

In the absence of precise forecasts and a comprehensive theory of weather, researchers cannot conclusively establish the effects of weather modification procedures such as cloud seeding. Any ensuing precipitation might have occurred without the procedure. And the amount of precipitation may have been more or less without the intervention—meteorologists presently have no way of being sure.

Weather balloons help researchers measure present conditions and forecast future weather. A researcher at the National Weather Service station in Topeka, Kansas, is releasing this weather balloon, which carries a radiosonde—an instrument that makes atmospheric measurements and transmits the data to a computer. *[AP Photo/The Lawrence Journal-World, Nick Krug]*

To tackle this problem, researchers turn to statistics. Statistics condense and summarize the values of repeated measurements, producing an average and an indication of how much the values vary from this average—the variability. Variability is important because it tells how often and by how much the value can be expected to deviate from normal. For example, suppose researchers measure the precipitation from clouds of a certain size and height, under certain conditions of temperature and humidity. The average precipitation of these clouds is, say, X, and virtually no cloud produces $2X$. Suppose a cloud seeding invokes $2X$ precipitation. The seeding may be responsible because this level of precipitation rarely occurs naturally. If this happens often enough, scientists become convinced that the modification procedure has an effect.

The problem with weather modification research is that weather can be extremely variable. Researchers also have to accept the conditions as they exist—they are not working in the laboratory where they can perform a number of trials with the same factors—which adds to the variability. As a result, weather modification researchers have trouble drawing firm conclusions from statistical analysis. The high degree of variability can be as great or greater than the effect, if any, of cloud seeding, making it much harder to discern if weather modification is an important factor or not.

Because of the potential benefits of weather modification technology, the National Academies, a body of four U.S. organizations that provide scientific advice to government agencies and policy-makers, have produced four reports on the subject—in 1964, 1966, 1973, and 2003. The gap between the third and fourth indicates that weather modification fell out of the spotlight for a while, in part because scientists who wrote the first three reports were skeptical. As described in the sidebar on page 154, the National Academies are influential in scientific matters and in the direction of, and funding for, research and development. But in 2000, the National Academies' Board on Atmospheric Sciences and Climate organized a meeting to take up the subject of weather modification once again, and in 2001 convened a committee—the Committee on the Status of and Future Directions in U.S. Weather Modification Research and Operations—of experts to sift through the evidence that had accumulated since the previous report. The University of Virginia researcher Michael Garstang chaired the committee.

The committee's 2003 report was also critical of weather modification, citing the problems of variability and the lack of a comprehensive understanding of atmospheric phenomena. But the committee recommended

National Academies

The National Academies are four organizations: the National Academy of Sciences, the National Academy of Engineering, the Institute of Medicine, and the National Research Council. The National Academy of Sciences is the oldest, originating during the Civil War. President Abraham Lincoln signed the Act of Incorporation on March 3, 1863, establishing the National Academy of Sciences and its 50 charter members to serve the national interests in matters of science and technology.

Advice on military technology predominated early efforts, but the broad scope of science and technology demands expertise from many different fields. In 1916, the National Academy of Sciences founded the National Research Council, which addresses specific research issues and enlists the help of experts at institutes and universities all over the country. The National Academies also broadened their scope by establishing the National Academy of Engineering in 1964 and the Institute of Medicine in 1970.

Membership in these National Academies is a mark of great achievement. Current members of these organizations elect new members based on records of accomplishment in their specialties. Members are not full-time employees but instead perform their services to the academy in addition to their duties as professors or researchers at universities or institutes. The National Academy of Sciences, for example, consists of about 1,800 members who serve on or select members for specific committees, organize meetings and workshops, and analyze issues of national importance.

further research instead of dismissing the concept. In their report, *Critical Issues in Weather Modification Research*, they wrote, "The Committee concludes that there still is no convincing scientific proof of the efficacy of intentional weather modification efforts. In some instances there are strong indications of induced changes, but this evidence has not been subjected to tests of significance and reproducibility. This does not challenge the scientific basis of weather modification concepts. Rather it is the absence of adequate understanding of critical atmospheric processes that, in turn, lead to a failure in producing predictable, detectable, and verifiable results. Questions such as the transferability of seeding techniques or whether seeding in one location can reduce precipitation in other areas can only be addressed through sustained research of the underlying science combined with carefully crafted hypotheses and physical and statistical experiments."

Because of the skepticism of many scientists, little money has flowed from the federal government into weather modification research projects. The projects and operations in the United States have received support from states such as Utah and Nevada.

One recent weather modification project in Wyoming is attempting to address some of the questions of the skeptics. Seeding operations for the Wyoming Weather Modification Pilot Project, sponsored by the Wyoming Water Development Commission, began in 2006. The five-

A meteorologist in Colorado measures rainfall using a gauge protected by fences to prevent winds from skewing the measurement. *[Scott Bauer/U.S. Department of Agriculture/Photo Researchers, Inc.]*

year project is using air- and ground-based generators to seed winter clouds. In addition to contractors that perform the seeding operations, participants include scientists from the National Center for Atmospheric Research (NCAR), the University of Wyoming, the Desert Research Institute, and other organizations to design the experiments and measure the results. Researchers hope to collect data from up to 200 trials.

Seeding long-lived winter clouds to elicit snow is common in the dry western states. The snow accumulates, resulting in more water during spring thaws. Although states report increases of 10 or 15 percent, these yields are well within natural variability.

The Wyoming project scientists are tackling the variability problem by carefully seeding clouds over a certain mountain range and leaving alone the clouds of another range in the area. This provides two sets of data—one that reflects the input of the experimental procedure and another which does not. Data coming from conditions or objects that have not been subjected to experimental treatment are often called controls in laboratory research, and, although it is difficult to hold all other variables constant in field research, statistical analysis might be able to distinguish precipitation levels between the two data sets. Researchers are also examining the snow for traces of the seed nuclei (silver iodide), which suggests that the substance played a role in the formation of the crystals that eventually fell.

An important part of the project's design is that the scientists who are evaluating the results of the cloud seeding are independent, meaning that they are not involved or personally committed to commercial operations in weather modification. This removes the possibility of bias. When this project is completed, scientists will have gained a better idea of the effectiveness of winter cloud seeding.

Weather modification might also be applied for even greater potential benefits—reducing major storms such as hurricanes. But the stakes will be high. And an early project, called Stormfury, did not yield conclusive results.

PROJECT STORMFURY—AN EFFORT TO WEAKEN TROPICAL CYCLONES

As discussed in chapter 5, hurricanes need the heat from warm water to drive their swirling winds. Evaporation releases energy that results in

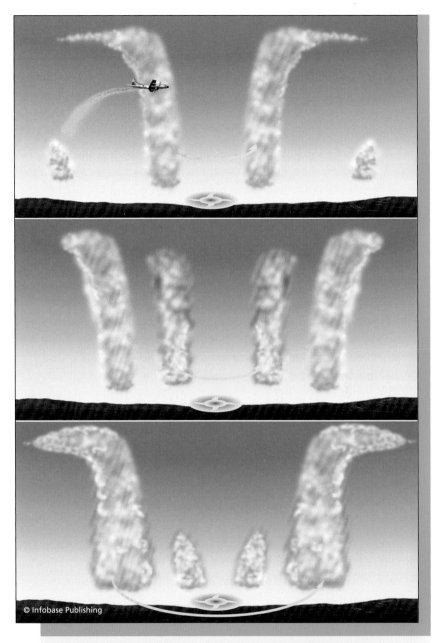

© Infobase Publishing

In the top panel of the figure, an airplane seeds clouds in and around the eyewall of a hurricane. If the operation generates precipitation and unstable air, as shown in the middle panel, then the old eyewall may weaken. The new eyewall may have a greater diameter, resulting in reduced wind speed.

temperature gradients and pressure differences, leading to winds that begin to spin around an area of low pressure. Severe thunderstorms and the strongest winds form the eyewall surrounding the calm eye. Hurricanes weaken rapidly when they leave tropical water and strike land or encounter cool water in the northern latitudes. As the winds of the cyclone reduce speed, the eye increases in radius and gradually disappears.

In the early 1960s, meteorologists hypothesized that the upper levels of the region around the eyewall might contain supercooled water. If so, then seeding of this region could generate precipitation and cause additional storms to crop up in and around the eyewall. These storms would rob some of the energy from the eyewall and perhaps result in the reformation, or migration, of the eyewall to a larger radius. The figure on page 157 illustrates this hypothesis. Because angular momentum is conserved, the increased radius would result in a lower wind speed (angular momentum is related to the product of size and angular speed, so an increase in radius must coincide with a decrease in speed if the product is constant). The most suitable kind of hurricane for this weather modification procedure has a well-defined eye and a sharp increase in wind speed near the eyewall.

Researchers from the U.S. Navy and the Weather Bureau tested this hypothesis on September 16, 1961, when a Navy airplane released canisters of silver iodide around the eyewall of Hurricane Esther. Observers noted some weakening of the storm following the seeding. On the following day, the airplane seeded again, but the canisters fell outside of the target zone, and no weakening was observed. The success of the first mission encouraged scientists to conduct additional research. In 1962, Project Stormfury was officially launched. Project Cirrus had dabbled in storm modification, but Project Stormfury was devoted to this field of research. Participants included the U.S. Navy, Air Force, Weather Bureau, and the National Hurricane Center.

No suitable hurricane presented itself in 1962, and researchers wanted to be careful not to conduct experiments on storms close to land in case the procedures yielded unexpected results. But Project Stormfury researchers targeted several hurricanes in the 1960s and early 1970s. Summarizing the project, the National Oceanic and Atmospheric Administration (NOAA) researcher Hugh Willoughby and his colleagues wrote in a 1985 issue of the *Bulletin of the American Meteorological Society,* "Modification was attempted in four hurricanes on eight different days. On four of these days, the winds decreased by between 10 and 30 percent. The lack of response on the other days was interpreted to be the result of faulty execution of the experiment or poorly selected subjects."

Researchers enjoyed particular success with Hurricane Debbie in 1969. As Willoughby and his colleagues noted, "The winds decreased by 31 percent and 15 percent respectively on the two seeding days. Moreover, the decreases occurred at the time expected if they were caused by seeding, and their magnitudes were believed to exceed most of those observed in unmodified hurricanes."

These promising results were encouraging. Yet Project Stormfury and its fundamental hypothesis began running into problems. The main idea held that the seeding triggering precipitation and instability that enlarged the eyewall and accordingly reduced wind speed. But in order to have an effect, cloud seeding generally requires supercooled water. As the data accumulated—including data gathered during Project Stormfury—meteorologists began to question the notion that much supercooled water was present in hurricanes. In addition, hurricane researchers discovered that renewal of the eyewall of these storms occurs naturally (see chapter 5).

Although Project Stormfury produced some apparent successes, scientists now had reason to doubt the effectiveness of cloud seeding in hurricanes. They also had an alternative explanation for the weakening—hurricanes and their eyewalls undergo routine variability and restructuring. This variability meant that researchers could not distinguish between weakening that resulted from cloud seeding and that from natural causes. As with rainmaking, skeptics could and did argue that the modification would have happened even in the absence of the procedure.

Hurricane Ginger, a 1971 hurricane, was the last hurricane to be seeded. This effort was unsuccessful, though Ginger was not a good candidate since it lacked a well-defined eye. Doubts about the effectiveness of hurricane seeding grew. Project Stormfury continued though mostly inactive, officially closing in 1982.

There has been little research into hurricane modification since Project Stormfury shut down. Some scientists have made speculative proposals, though nothing has come of them. For example, the meteorologist Daniel Rosenfeld of Hebrew University in Israel proposed in 2007 that dropping microscopic dust particles on hurricanes could weaken their force. Computer simulations showed that the dust could act to collect moisture, reducing rain and temperatures, and decreasing the supply of energy that fuels hurricanes.

In 2008, the U.S. Department of Homeland Security sponsored a workshop at the NOAA Earth System Research Laboratory in Boulder, Colorado, to discuss the possibility of hurricane modification. Considering the billions of dollars of destruction and the loss of life that Hurricane Katrina and other major storms cause, renewal of a research program into hurricane weakening might pay great dividends.

But NOAA does not appear enthusiastic about this research and has yet to make any plans for hurricane modification projects. Forecasting is the main focus of current research on hurricanes. Chapter 5 described the efforts of NOAA and other weather researchers to improve forecasts of the path and intensity of storms as they near land. These researchers believe that storm prediction methods offer one of the best chances of reducing the hazards associated with hurricanes.

A reluctance to revive hurricane modification efforts may be due to several things. One reason is that these campaigns are expensive and the theoretical basis for hurricane modification is not yet fully established. Another reason may have to do with the fear of unintended consequences. As hurricane forecasters know well, these storms do not always behave as expected.

UNINTENDED CONSEQUENCES OF WEATHER MODIFICATION

The complexity that makes weather systems difficult to predict also makes them difficult to control. Since meteorologists' knowledge is not complete, a modification scheme could easily produce unanticipated effects.

For example, suppose a hurricane intensified after a modification attempt or suddenly veered to another course. Although it would be difficult to prove that the modification caused the intensification (for the same reasons that proof of the effectiveness of cloud seeding is difficult), affected residents might be suspicious. And if the storm unexpectedly turned after the modification and struck an unprepared area of the coast, the residents might be caught off guard. The weather modification operation may or may not have been responsible, but the operators would likely be in court defending themselves from multiple lawsuits. Fear of lawsuits was one of the reasons why General Electric Research Laboratory wanted the government involved in its early projects in the 1940s.

The possibility of unintended consequences raises questions about the viability of any sort of ambitious weather modification project. Seeding winter clouds over deserted mountain ranges causes few concerns, but interfering with the evolution of a major hurricane or other storm systems could be risky.

Even rainmaking projects, if carried over a broad region or for an extended period of time, pose certain issues. If successful, cloud seeding wrings some of the moisture out of the air as it flows past a certain area. But this moisture could have fallen somewhere else if it had been allowed to accumulate. By squeezing this water out of the clouds, weather modification programs may be benefiting the area of operation at the expense of other areas downwind. Since water is a scarce resource, particularly in the West, pervasive cloud seeding in one area may have a negative impact on another—and the people living in that area may consider hiring an attorney. Water rights and allocation laws are complicated and depend on the jurisdiction. The legal issues of widespread weather modification may be difficult to resolve.

Another issue is the possibility of deliberately creating negative outcomes with weather modification operations in order to suppress a region or country. Such weather "warfare" is frightening because it could be potentially devastating. But large-scale operations seem remote at the present time given the incipient state of the technology, and the complexity of weather systems may forever preclude hostile modifications—the aggressors may put themselves in just as much risk as the intended victims.

Attempts to control weather must also take into account global effects that might arise. For example, hurricanes are nature's way of transferring energy from the warm tropical regions to cooler parts of the world. If people began interfering with this process, no one knows what the consequences could be for Earth's climate. This issue merits consideration if and when officials begin to discuss serious weather modification programs.

CONCLUSION

Although the possibility of unintended consequences warrants caution, meteorologists are making progress in the study of weather systems and the factors that can influence or alter their course of development. Evaluating the effectiveness of weather modification techniques is a priority. Researchers in Wyoming and elsewhere are in the process of accumulating data that may clearly prove the point—or prove that the skeptics were right.

The future of weather modification depends to a certain extent on the effectiveness of cloud seeding, which is the most common and well-developed method. But researchers at the frontiers of weather and climate science have their eye on other goals. One of the most important of these goals involves global climate change.

Human activity and technology have drastically altered Earth's landscape as farms and cities spread out over much of the land. Waste products and the production and release of a number of substances, including GHGs, have affected the planet's composition, particularly its atmosphere. But if human activity is changing weather and climate in unintended ways, some people wonder if human activity, such as weather modification, might be able to counter these undesirable effects.

The risk of further unintended consequences must figure into any concerted plan to reverse or counter global climate change. But as meteorological knowledge grows, researchers may become more confident about reducing the uncertainty. Considering the scale of the problem and the potential need for action that could have a strongly negative impact on the world's economy, alternative options should be considered, including weather and climate modification. The concluding section of this book discusses some of the strategies that researchers are studying.

Other visionary projects center around cloud-seeding techniques applied on a nationwide scale and in a more controlled manner. Even if the current methods of cloud seeding do not prove to be highly effective, laboratory and field experiments demonstrate that clouds can be influenced—ever since the early work of Schaefer and Vonnegut researchers have known that condensation nuclei have an effect, though not necessarily the desired one. Refinements of the present techniques may be necessary, but if an effective technique exists or can be developed, cloud seeding could provide the means to control certain aspects of the weather.

Imagine, for example, a network of sensors scattered throughout the United States. Surface sensors could measure humidity, winds, temperature, and air pressure and transmit the reports to a central computer. Sensors in weather balloons provide the same data at various altitudes. Images and data from satellites complete the picture, showing the distribution of clouds.

These sensors and images would generate an enormous amount of data. A tremendously fast computer or network of computers must analyze this data in real-time—as the situation develops. The programs monitor current weather conditions and simulate or predict what will happen in the near future, perhaps over the next few hours. Any undesirable weather patterns might be averted by the application of appropriate cloud-seeding generators at the right moment and in the right area. If, say, a network of these generators were installed and available, the computer could automatically direct their operation.

The present state of weather modification science and technology is not sophisticated enough to construct an automated system that could control the weather. But researchers may arrive at this point in the next few decades if this work receives attention and funding. An editorial published in *Nature* in 2008 promoted this research: "The stakes are high, as weather modification is one of those areas in which science can have an immediate and obvious benefit for society. It's long past time to invest modest funds in the basic understanding of it." Scientists at the frontier of weather and climate research will be working to make this happen as soon as possible.

CHRONOLOGY

1750s   American statesman, businessman, and inventor Benjamin Franklin (1706–90) develops the lightning rod.

1891–92   The Department of Agriculture enlists the American attorney and engineer Robert G. Dyrenforth

(1844–1910) to test the idea that loud noises can produce rain.

1946　The General Electric Research Laboratory scientists Vincent Schaefer (1906–93) and Bernard Vonnegut (1914–97) discover that certain substances such as dry ice and silver iodide can act as condensation nuclei.

1947–52　Project Cirrus conducts cloud-seeding experiments.

1950s　Cloud-seeding projects begin in Utah.

1961　Researchers from the U.S. Navy and the Weather Bureau use silver iodide in the attempt to weaken Hurricane Esther. The effort appears partially successful.

1962–82　Project Stormfury investigates the possibility of weakening hurricanes by cloud seeding.

1964　The National Academies issue their first report on weather modification. The report sounds a skeptical note.

1973　The Utah legislature passes the Utah Cloud Seeding Act, which makes provisions for licensing cloud-seeding operators and allows the Utah Division of Water Resources to sponsor projects.

1980s–90s　Agencies of some states in the United States, particular in the West, fund various cloud-seeding projects, federal government involvement is virtually nil.

2003　The National Academies issue their fourth report, *Critical Issues in Weather Modification Research*, on weather modification. Although highlighting problems with interpretation and measurement, the report recommends further research.

2006   Field tests in the Wyoming Weather Modifica-
                tion Pilot Project begin. Participants include sci-
                entists from the National Center for Atmospheric
                Research (NCAR), the University of Wyoming,
                the Desert Research Institute, and other organiza-
                tions to design the experiments and measure the
                results.

2008   Chinese meteorologists attempt to keep the rain at
                bay during the Beijing Olympics, with mixed results.

                The Department of Homeland Security sponsors
                a workshop at the NOAA Earth System Research
                Laboratory in Boulder, Colorado, to discuss re-
                newed research into hurricane modification.

2009   The North American Interstate Weather Modifi-
                cation Council lists dozens of ongoing projects in
                10 states in the United States and one Canadian
                province.

FURTHER RESOURCES
Print and Internet

Casper Star-Tribune Online. "In Its Fourth Year, Wyoming's $8.8 Million
   Cloud-Seeding Experiment Is Drawing Big-Time Attention" (2/14/09).
   Available online. URL: http://casperstartribune.net/articles/2009/
   02/15/news/wyoming/4a8a6b0fc02891628725755d00269578.txt.
   Accessed July 1, 2009. Staff writer Wes Smalling reports on the Wyo-
   ming Weather Modification Pilot Project.

Committee on the Status of and Future Directions in U.S. Weather
   Modification Research and Operations. *Critical Issues in Weather
   Modification Research.* Washington, D.C.: National Academies
   Press, 2003. The National Academies' latest report on weather modi-
   fication analyzes the current state of research and discusses options
   and opportunities for future research projects.

Egan, Timothy. *The Worst Hard Time: The Untold Story of Those Who Survived the Great American Dust Bowl.* New York: Mariner Books, 2006. This history of the 1930s dust bowl describes the economic, ecological, and human catastrophe in vivid detail.

Franklin Institute. "The Lightning Rod." Available online. URL: http://www.fi.edu/pieces/hongell/. Accessed July 1, 2009. This article briefly describes Benjamin Franklin's invention of the lightning rod.

General Electric Research Laboratory. *History of Project Cirrus* (July 1952). Available online. URL: http://handle.dtic.mil/100.2/AD006880. Accessed July 1, 2009. This document describes the techniques and results of Project Cirrus, an early series of cloud-seeding experiments.

Nature editorial. "Change in the Weather" (6/18/08). Available online. URL: http://www.nature.com/nature/journal/v453/n7198/full/453957b.html. Accessed July 1, 2009. This editorial calls for more studies of weather modification.

O'Hanlon, Larry. "Weather Modification Comes of Age" (4/22/08). Available online. URL: http://dsc.discovery.com/news/2008/04/22/cloud-seeding-weather.html. Accessed July 1, 2009. This short article reports the mixed feelings of meteorologists concerning weather modification procedures.

Outwater, Alice. *Water: A Natural History.* New York: Basic Books, 1996. Water is constantly on the go. The author eloquently describes water's journey from lake to house drain and back again as it travels through complex ecological systems.

Pretor-Pinney, Gavin. *The Cloudspotter's Guide: The Science, History, and Culture of Clouds.* New York: Perigee, 2006. This book offers an entertaining look at how clouds form and the history of cloud watching as a science and a pastime.

Texas Council for the Humanities Resource Center. "The Dust Bowl." Available online. URL: http://www.humanities-interactive.org/texas/dustbowl/. Accessed July 1, 2009. The hardships of life in the dust bowl are highlighted, including many photographs and an essay.

Willoughby, H. E., D. P. Jorgensen, R. A. Black, and S. L. Rosenthal. "Project STORMFURY: A Scientific Chronicle 1962–1983." *Bulletin of the American Meteorological Society* 66 (1985): 505–514. The authors discuss the rationale and history of the project that studied the possibility of weakening hurricanes.

Web Sites

National Integrated Drought Information System (NIDIS) home page. Available online. URL: http://www.drought.gov. Accessed July 1, 2009. The NIDIS Web site offers maps and information showing which parts of the United States are currently experiencing a drought and how long it might last.

North American Interstate Weather Modification Council. Available online. URL: http://www.naiwmc.org/. Accessed July 1, 2009. This organization is composed of researchers and regulatory agencies involved in weather modification. The Web site contains information on cloud-seeding technology and projects.

Utah Cloud Seeding. Available online. URL: http://www.water.utah. gov/cloudseeding/default.asp. Accessed July 1, 2009. The Utah Division of Water Resources has been involved in a number of cloud-seeding projects. This Web site describes the technology and the results.

FINAL THOUGHTS

Much of the first three chapters of this book involved research on global climate change. Scientists have documented the general rising of surface temperatures and melting of glaciers and Arctic sea ice. According to many scientists and the IPCC, human activities such as greenhouse gas emissions are the main cause of these changes. Human construction, industry, and technology have altered Earth in a variety of ways, some for the better and some for the worse.

Scientists do not yet fully understand global climate change and its consequences. Predictions of the problems that loom in the near future—additional increases in average temperatures, disrupted weather patterns, and rising sea levels—range from mild to catastrophic, but almost everyone agrees that humans face a certain number of climate challenges ahead.

The topics in several chapters of this book directly touch upon modeling and forecasting efforts. A better understanding of weather and climate will help scientists predict what further changes will occur and what is causing them. Countering global climate change will probably require a concerted effort to eliminate or reduce the activities that are primarily responsible. People can reduce greenhouse gas emissions, for example, by cutting down on the use of fossil fuels—coal, oil, and natural gas—the combustion of which generates carbon dioxide and other greenhouse gases. But fossil fuel consumption will not stop entirely any time in the near future. As a cheap source of energy, fossil fuels will continue to be an important component of the world's economy.

A supplemental approach to fighting global climate change is to counteract the changes by some sort of artificial means. This strategy can be called climate modification. Two active areas of research involve the enhancement of carbon sinks and engineered changes in the planet's reflectivity or albedo.

A carbon sink is a reservoir or deposit of carbon, usually in the form of carbon-containing compounds. The carbon stored in these reservoirs is generally unavailable to contribute to the increasing amount of carbon dioxide, methane, and other greenhouse gases. Natural carbon sinks include forests, soils, and the ocean. For example, plants absorb carbon dioxide from the atmosphere and incorporate the carbon into various organic compounds such as carbohydrates. This process removes carbon dioxide from the air and locks it into other substances. Although much of this carbon gets recycled into the atmosphere again, the reservoir holds a large quantity at any given time. When plants die, some of this carbon gets buried and removed from circulation. Oceans are particularly large carbon sinks. Water absorbs carbon dioxide, which gets stored as carbonate in the shells

Human activities, technology, and metropolitan areas have altered Earth's landscape and environment—part of Tokyo, Japan, is shown here. *(James B. Adson/Shutterstock)*

of certain marine organisms or taken up by photosynthetic plankton. Some of the carbon ends up deposited as sediment.

To enhance the natural carbon sinks, some researchers are working on schemes to capture and store carbon dioxide emissions. For example, utilities powered with fossil fuels tend to generate huge amounts of carbon dioxide. Shutting down all industrial processes that emit greenhouse gases such as carbon dioxide is not realistic—the economy is presently too dependent on these processes—but a reduction in emissions would be beneficial. In addition to scaling down these processes, people can reduce emissions by capturing some of the carbon dioxide from the exhaust gases and depositing it into a carbon sink. Chemicals that absorb or react with carbon dioxide may be able to scrub much of the carbon dioxide from the waste.

Where to deposit this carbon dioxide is an important issue. The sink needs to lie undisturbed in order to remove the carbon from circulation. Options include abandoned oil fields or deep saltwater aquifers (underground water reservoirs) far below the surface. The idea is to pump the carbon dioxide into the sinks and let it settle or dissolve. Sinks must be below a layer of dense rock that prevents the gas from leaking out. If these technologies are developed and implemented, carbon dioxide emissions can be reduced immediately, and civilization will have more time to wean itself from fossil fuels.

Another area of research revolves around increasing Earth's overall albedo. An object with a high albedo reflects much solar radiation. Since little of the energy is absorbed, the object's temperature does not rise too much. White surfaces, for example, tend to be cooler in the summer than dark ones because they absorb fewer of the Sun's rays.

Global temperatures have dropped briefly following powerful volcanic eruptions that spewed gas and ash into the atmosphere. The eruption of the Tambora Volcano in Indonesia in 1815 was one of the largest eruptions in history, and the following year was exceptionally cool—the "year without summer." Decreasing the incoming radiation by reflecting more of it into space is a viable method of cooling the planet.

One way of increasing reflectivity is to increase cloud cover. John Latham, a researcher at the Mesoscale and Microscale Meteorology Division at the National Center for Atmospheric Research, has proposed to seed clouds over the oceans. As described in chapter 6, cloud-seeding operations distribute condensation nuclei in the effort to elicit snow or rain, but in Latham's proposal, the goal is to boost the number of clouds, thereby reflecting more solar radiation into space. Future changes in cloud production and distribution are some of the most important unknowns in climate models, and clouds certainly have an effect on temperature, although no one knows for sure if Latham's idea would work.

Modification strategies such as Latham's proposal invoke a certain amount of controversy because some people believe that society should be more focused on reducing emissions than on introducing additional modifications. And no one knows if these strategies will be necessary or would be effective if attempted. But researchers can test their ideas on a small scale or in the laboratory and refine the methods that appear promising. In the opinion of researchers such as Latham and many of his colleagues, society should not skimp on research for a problem of the magnitude of global climate change. Scientists at the frontier of weather and climate research may one day be called upon to exercise an unprecedented level of control.

GLOSSARY

aerosols   suspensions of small particles or droplets in air

albedo   the fraction of light an object reflects, which is a measure of its surface reflectivity

butterfly effect   an extreme sensitivity to conditions, such as exhibited by weather

cirrus clouds   masses of visible droplets or crystals in the atmosphere characterized by a thin, wispy appearance

climate forcings   mechanisms that cause or force the climate to change

condensation nucleus   a particle around which water vapor transitions to the liquid phase

cryosphere   regions containing snow and ice, such as sea ice, ice sheets and caps, glaciers, snow fields, and permafrost

cumulus clouds   masses of visible droplets or crystals in the atmosphere with well-defined edges and a puffy appearance

dendroclimatology   the study of trees in relation to climate

drought   an extended period of dry conditions

eccentricity   in orbits, a measure of shape, with lower values representing circular orbits and higher values representing flattened or elliptical orbits

electromagnetic radiation   electric and magnetic field oscillations of various frequencies that propagate in space or through matter and carry energy

electron   negatively charged particle and component of an atom, where it swarms around the nucleus

eyewall   region surrounding the calm center of a hurricane and containing the strongest winds

fronts   boundaries between two air masses having different densities, which are the sites of much meteorological activity, such as storms

glaciers   large bodies of slowly moving ice

greenhouse gases   gaseous substances that absorb electromagnetic radiation in the infrared range and tend to trap heat when present in the atmosphere

half-life   the amount of time required for one-half of a radioactive substance to decay

humidity   amount of water vapor in air

isotopes   atoms of the same element but with different numbers of neutrons

landspout   a tornado occurring on land but not associated with a supercell

mesocyclone   a well-organized vortex within a thunderstorm

meteorology   the study of atmospheric phenomena and processes such as weather

NASA   *See* **National Aeronautics and Space Administration**

National Aeronautics and Space Administration   the U.S. agency devoted to space science and exploration

National Oceanic and Atmospheric Administration   the U.S. agency devoted to the study and management of marine environments and resources as well as the atmosphere, weather, and climate

neutron electrically neutral particle, usually found in the nucleus of an atom

NOAA *See* **National Oceanic and Atmospheric Administration**

orographic clouds masses of visible droplets or crystals in the atmosphere that form when moist air moves over mountainous terrain and begins to rise

permafrost soil that is continually frozen

photosphere outer layer of the Sun

pollen sticky powder, made and released by certain plants, which contains grains consisting of reproductive cells

precipitation the fall of condensed water vapor in any form, such as rain, snow, ice, sleet, or hail, from the sky

proton positively charged particle, usually found in the nucleus of an atom

proxy climate data information collected from tree rings, ice cores, and other objects that constitute records of temperature, rainfall, and other important variables

radiometer an instrument to measure the amount of electromagnetic radiation

sea ice frozen ocean water that floats on the surface

sediments substances such as sand, mud, and organic material that settle to the bottom of a body of water

solar constant the average amount of radiation Earth receives from the Sun

solar irradiance amount of the Sun's radiation that arrives at a surface, such as the surface of Earth or its upper atmosphere

squall line a thunderstorm containing multiple cells or updrafts that form a line

summer solstice the first day of summer and the longest day of the year, when the Sun's position in the sky reaches its highest point of the year

sunspots dark regions on the Sun's disk that are at a lower temperature than the surrounding area

supercell a type of thunderstorm that generates a mesocyclone

supercooled water liquid at a temperature below the freezing point

thunderstorm a violent weather system exhibiting lightning discharges

tropical cyclone a storm characterized by winds whirling around an area of low air pressure that forms over warm water

tundra cold, treeless plains

vortex a spinning mass of air

water cycle the movement of water above, on, and under Earth as it evaporates from a pond or the ocean and then falls as precipitation, settling into a body of water and repeating the process

waterspout tornado that occurs over the water

water vapor gaseous phase of H_2O

wind shear a situation in which wind speed or direction changes across a given area, such as an increase in speed with increasing altitude

winter solstice the first day of winter and the shortest day of the year, when the noontime position of the Sun is lower than any other day

FURTHER RESOURCES

Print and Internet

Burt, Christopher. *Extreme Weather: A Guide and Record Book.* New York: W. W. Norton, 2007. Records, trends, historical facts, and descriptions of unusual weather events highlight this book.

Cox, John D. *Storm Watchers: The Turbulent History of Weather Prediction from Franklin's Kite to El Niño.* New York: Wiley, 2002. Radar, satellites, and sophisticated computer models have greatly improved weather forecasting techniques recently, but early meteorologists relied on a number of tools and observations to predict the weather. This book discusses the evolution of the science.

Dessler, Andrew E., and Edward A. Parson. *The Science and Politics of Global Climate Change: A Guide to the Debate.* Cambridge: Cambridge University Press, 2006. The authors cover the debate between scientists and policy-makers over the uncertain future of Earth's climate.

Dunlop, Storm. *The Weather Identification Handbook.* Guilford, Conn.: The Lyons Press, 2003. This book offers an illustrated guide to various weather and atmospheric phenomena such as cloud formations, weather maps, precipitation, storms, and many others.

Houghton, John. *Global Warming: The Complete Briefing,* 4th ed. Cambridge: Cambridge University Press, 2009. This text, intended for beginning college students, thoroughly explains the science of global climate change.

Lomborg, Bjorn. *Cool It: The Skeptical Environmentalist's Guide to Global Warming.* New York: Vintage, 2008. Lomborg, an outspoken economist and political scientist who does not shrink from taking controversial positions, presents a contrarian view that fears of global warming are overblown.

Mogil, H. Michael. *Extreme Weather: Understanding the Science of Hurricanes, Tornadoes, Floods, Heat Waves, Snow Storms, Global Warming and Other Atmospheric Disturbances.* New York: Black Dog and Leventhal Publishers, 2007. Well-illustrated and comprehensive, this book explains severe weather systems and how they develop.

Williams, Jack. *The Weather Book,* 2nd ed. New York: Vintage Books, 1997. Williams, one of the founding editors of the weather section in the newspaper *USA Today,* describes and explains weather systems with the aid of many colorful illustrations. Chapter topics include tornadoes, floods, snow and ice, hurricanes, and others.

Web Sites

Exploratorium. Available online. URL: http://www.exploratorium.edu/. Accessed July 1, 2009. The Exploratorium, a museum of science, art and human perception in San Francisco, has a fantastic Web site full of virtual exhibits, articles, and animations, including much of interest to weather and climate scientists.

How Stuff Works. Available online. URL: http://www.howstuffworks. com/. Accessed July 1, 2009. This Web site hosts a huge number of articles on all aspects of technology and science, including weather and climate.

National Aeronautics and Space Administration. Available online. URL: http://www.nasa.gov. Accessed July 1, 2009. NASA's Web site contains a huge amount of information on astronomy, physics, and earth science, and includes news and videos of NASA's many exciting projects.

National Center for Atmospheric Research. Available online. URL: http://www.ncar.ucar.edu/. Accessed July 1, 2009. NCAR supports a variety of research projects involving weather and climate. Their Web site provides news and information on the latest findings.

National Oceanic and Atmospheric Administration. Available online. URL: http://www.noaa.gov/. Accessed July 1, 2009. NOAA's Web site provides a huge amount of information on weather and climate issues as well as marine science and oceanography. Topics include basic information, observational data, and NOAA research.

National Weather Service. Available online. URL: http://www.nws. noaa.gov/. Accessed July 1, 2009. NWS's Web site contains warnings and forecasts, maps, radar observations, satellite and climate data, and information on many different aspects of weather and climate.

Online Meteorology Guides. Available online. URL: http://ww2010. atmos.uiuc.edu/(Gh)/guides/mtr/home.rxml. Accessed July 1, 2009. Hosted by the University of Illinois Urbana-Champaign, the topics for this well-illustrated set of educational modules include light and optics, clouds and precipitation, forces and winds, air masses and fronts, weather forecasting, severe storms, hurricanes, El Niño, and the water cycle.

Science*Daily.* Available online. URL: http://www.sciencedaily.com/. Accessed July 1, 2009. An excellent source for the latest research news, Science*Daily* posts hundreds of articles on all aspects of science. The articles are usually taken from press releases issued by the researcher's institution or by the journal that published the research. Main categories include Earth & Climate, Matter & Energy, Space & Time, and others.

Weather Channel. Available online. URL: http://www.weather.com/. Accessed July 1, 2009. The online presence of the Weather Channel contains current weather conditions, maps, radar reports, images, and articles on weather research projects.

INDEX